SpringerBriefs in Computer Science

Series Editors

Stan Zdonik
Peng Ning
Shashi Shekhar
Jonathan Katz
Xindong Wu
Lakhmi C. Jain
David Padua
Xuemin Shen
Borko Furht
V. S. Subrahmanian
Martial Hebert
Katsushi Ikeuchi
Bruno Siciliano

For further volumes:
http://www.springer.com/series/10028

Michal Forišek · Monika Steinová

Explaining Algorithms Using Metaphors

 Springer

Michal Forišek
Department of Computer Science
Comenius University
Bratislava
Slovakia

Monika Steinová
Department of Computer Science
ETH Zürich
Zurich
Switzerland

ISSN 2191-5768 ISSN 2191-5776 (electronic)
ISBN 978-1-4471-5018-3 ISBN 978-1-4471-5019-0 (eBook)
DOI 10.1007/978-1-4471-5019-0
Springer London Heidelberg New York Dordrecht

Library of Congress Control Number: 2013932561

© The Author(s) 2013
This work is subject to copyright. All rights are reserved by the Publisher, whether the whole or part of the material is concerned, specifically the rights of translation, reprinting, reuse of illustrations, recitation, broadcasting, reproduction on microfilms or in any other physical way, and transmission or information storage and retrieval, electronic adaptation, computer software, or by similar or dissimilar methodology now known or hereafter developed. Exempted from this legal reservation are brief excerpts in connection with reviews or scholarly analysis or material supplied specifically for the purpose of being entered and executed on a computer system, for exclusive use by the purchaser of the work. Duplication of this publication or parts thereof is permitted only under the provisions of the Copyright Law of the Publisher's location, in its current version, and permission for use must always be obtained from Springer. Permissions for use may be obtained through RightsLink at the Copyright Clearance Center. Violations are liable to prosecution under the respective Copyright Law.
The use of general descriptive names, registered names, trademarks, service marks, etc. in this publication does not imply, even in the absence of a specific statement, that such names are exempt from the relevant protective laws and regulations and therefore free for general use.
While the advice and information in this book are believed to be true and accurate at the date of publication, neither the authors nor the editors nor the publisher can accept any legal responsibility for any errors or omissions that may be made. The publisher makes no warranty, express or implied, with respect to the material contained herein.

Printed on acid-free paper

Springer is part of Springer Science+Business Media (www.springer.com)

Preface

Motivation

> Feynman was a truly great teacher. He prided himself on being able to devise ways to explain even the most profound ideas to beginning students. Once, I said to him, "Dick, explain to me, so that I can understand it, why spin one-half particles obey Fermi-Dirac statistics." Sizing up his audience perfectly, Feynman said, "I'll prepare a freshman lecture on it." But he came back a few days later to say, "I couldn't do it. I couldn't reduce it to the freshman level. That means we don't really understand it."
>
> —David L. Goodstein

The above quotation nicely expresses our feelings about education. It is our opinion that the ability to explain a subject is an indicator of having the deepest understanding of the subject. Without knowing the subject like the back of one's hand, it is impossible to explain it in (relatively) simple terms. This is true for all subjects, but perhaps most significant when the subject is scientific.

In science, the research does sometimes bring a new discovery. The discovery then initiates a wave of new results—both pushing the boundary further and filling in the blanks around the initial discovery. But even once all the blanks have been filled in, it still makes sense to continue doing the research. The goal of this phase is to simplify the results as much as possible: to look for a more systematic approach, to find the most concise way of describing and presenting them, to find the simplest possible proof... In other words, to extract the essence. This phase of research has two major benefits: First, there is the pure scientific benefit of understanding the subject better, which often allows the researchers to discover patterns that were invisible in the first results, or to generalize or otherwise improve the results. The second benefit is for the sake of education. The better we understand what's going on, the easier we can explain it to the next generation of researchers, the faster the research can move forward.

The history of science abounds with examples where a new, better way of looking at things significantly facilitated our progress. For instance, Euclidean geometry was almost without any progress for centuries, until René Descartes came up with the coordinate system. This discovery gave birth to analytic geometry—and suddenly, we saw connections between geometry, algebra, and number theory. And only with this connection it was later possible to prove that it is not possible to trisect a general angle using just a compass and a straightedge. Later came another significant step: the complex numbers. It turned out that in analytic geometry they can be used to represent angles and rotations easily. And in fact, complex numbers (and the more general quaternions) are still frequently used in modern computer graphics for that very purpose.

The above example is just one of our favorites, but we could have easily picked hundreds of others. They are all consequences of a higher principle: *Our thinking is limited by the language we use.* A more powerful language allows us to be more efficient when dealing with the stuff we already know, and at the same time it helps us grasp and investigate systems that were too complex before.

> Hunagarian mathematician Paul Erdős, although an atheist, spoke of an imaginary book, in which God has written down all the most beautiful mathematical proofs. When Erdős wanted to express particular appreciation of a proof, he would exclaim "This one's from the Book!"
> —John Francis

We fully subscribe to this belief. In fact, being computer scientists, we like to push it a bit further: There has to be a second volume to the book—one with all the most beautiful algorithms. Sadly, we also understand why Erdős assigns the authorship to a supernatural entity—writing such a book has to be a task that can never be finished by mere mortals. And our second volume would be even harder to write.

Many researchers (especially mathematicians) view computer science just as one of the many areas of mathematics. We mostly agree with this point of view. However, the field of computer science has some unique features. One of them is very prominent in algorithm design.

In mathematics, there is no such thing as a better or a worse proof. Once you have a proof, the corresponding theorem is proved. Of course, some proofs are short, simple, and illuminating, while others are long, clumsy, filled with special cases, and downright ugly. But a proof is a proof, and when showing that a theorem holds, any proof will do.[1]

But once we move from proving mathematical theorems to designing algorithms, we encounter a whole new dimension: *efficiency*. Not only can an algorithm be nice or ugly, simple or convoluted, we must also take its time complexity into account. Often, finding an algorithm that solves a particular problem is not the end of the road—it may still be possible to find another one that is strictly better.[2]

---

[1] Actually, that's not *entirely* true. For instance, it makes sense to distinguish between constructive and non-constructive proofs.

[2] What does *better* mean? Usually, the asymptotic behavior of the time complexity is what makes an algorithm better. However, there are also many situations in practice where actual execution speed and/or code simplicity matters.

And make no mistake—while there is a strong positive correlation, *beauty* and *efficiency* are still two very different aspects. For instance, consider the Floyd-Warshall algorithm that computes all-pairs shortest paths in a graph:

```
// Floyd-Warshall algorithm:
for i in 1..n:
  for j in 1..n:
    distance[i][j] = edge_length[i][j]   (zero if i=j, inf. if no such edge)
for k in 1..n:
  for i in 1..n:
    for j in 1..n:
      distance[i][j] = min( distance[i][j], distance[i][k] + distance[k][j] )
```

To most of the computer scientists we know, the sheer minimalistic simplicity makes this one of the most beautiful algorithms. Of course, nowadays we already know multiple algorithms that solve the same problem more efficiently—for instance, algorithms that reduce it to special matrix multiplication, and even a randomized $O(n^2 \log n)$ algorithm due to Moffat and Takaoka. But is any of them more beautiful? We leave the decision to you.

Where this Book Comes in

This book describes multiple algorithms. However, it does not aim to be an algorithm textbook.

The book should be readable to a university student interested in algorithms. However, the said students are not our primary audience—their teachers are.

> Teaching is the only major occupation of man for which we have not yet developed tools that make an average person capable of competence and performance. In teaching we rely on the "naturals", the ones who somehow know how to teach.
> — Peter F. Drucker

No book will turn a layman into an excellent teacher. But, dear reader, if you are a computer science teacher and you are anything like us, you probably already spent months of your life hunting for resources that would help you be better. It would make us happy if this book turns out to be a helpful resource to you, and helps you a little bit in this life-long quest.

It would be preposterous to claim that our book is an excerpt from "The Book", but we did our best to achieve some kind of a resemblance. Only time will tell how close we made it.

Acknowledgments

Parts of this book were originally published as the authors' research paper titled "Metaphors and Analogies for Teaching Algorithms" on the conference SIGCSE 2012.

This book would not be here at all, were it not for the community of people involved in organizing programming contests in Slovakia. The members of this community are students and faculty members of Comenius University in Bratislava, Slovakia.

The most significant international activity of this community is the preparation of the annual *Internet Problem Solving Contest* (IPSC, http://ipsc.ksp.sk/)—an online competition that tries to push the boundary of, traditional programming contests. A total of 1306 teams from 81 countries have registered for IPSC 2012.

In addition to IPSC, our community is responsible for running the most significant national programming competitions for secondary school students—the Olympiad in Informatics and the Correspondence Seminar in Programming. Twice a year we organize a camp for the best and brightest secondary school students. Most of the materials presented in this book were originally developed for these camps, and later also used in university courses.

We would never be able to get where we are in our lives, were it not for this community. For that, and for all the help with testing and tweaking some of the materials in this book, we are eternally grateful.

Bratislava, Slovakia Michal Forišek
Zurich, Switzerland Monika Steinová
December 2012

Contents

1 Introduction 1
   1.1 Metaphors in Education 1
        1.1.1 Terminology 1
        1.1.2 Metaphor as a Teaching Tool 3
   1.2 Metaphors and Computers 7
   1.3 How to Read the Main Chapters 8
   References 9

2 Graph Algorithms 11
   2.1 Single-Source Shortest Paths in Graphs 11
        2.1.1 Overview 11
        2.1.2 Metaphor 13
        2.1.3 Analysis 18
        2.1.4 Experience 19
        2.1.5 Exercises 19
   2.2 Longest Paths in Trees 20
        2.2.1 Overview 20
        2.2.2 Metaphor 21
        2.2.3 Analysis 26
        2.2.4 Experience 27
        2.2.5 Exercises 28
   References 28

3 Computational Geometry 31
   3.1 Shortest Path with Obstacles 31
        3.1.1 Overview 31
        3.1.2 Metaphor 32
        3.1.3 Analysis 34
        3.1.4 Experience 36
        3.1.5 Exercises 36
   3.2 Distance Between Line Segments 37
        3.2.1 Overview 37
        3.2.2 Metaphor 38

		3.2.3 Analysis	41
		3.2.4 Experience	43
		3.2.5 Exercises	44
	3.3	Winding Number	44
		3.3.1 Overview	44
		3.3.2 Metaphor	46
		3.3.3 Analysis	48
		3.3.4 Experience	50
		3.3.5 Exercises	50
	3.4	Polygon Triangulation	51
		3.4.1 Overview	51
		3.4.2 Metaphor	54
		3.4.3 Analysis	54
		3.4.4 Experience	56
		3.4.5 Exercises	56
	References	57
4	**Strings and Sequences**	59
	4.1	Stacks and Queues	59
		4.1.1 Overview	59
		4.1.2 Metaphor	60
		4.1.3 Analysis	61
		4.1.4 Experience	61
		4.1.5 Exercises	62
	4.2	Median as the Optimal Meeting Spot	62
		4.2.1 Overview	62
		4.2.2 Metaphor	63
		4.2.3 Analysis	64
		4.2.4 Experience	65
		4.2.5 Exercises	65
	4.3	Substring Search	66
		4.3.1 Overview	66
		4.3.2 Metaphor	67
		4.3.3 Analysis	76
		4.3.4 Experience	76
		4.3.5 Exercises	77
	References	78
Solutions to Exercises	79
Index	93

Chapter 1
Introduction

1.1 Metaphors in Education

As you probably guessed from the title of this book, we will be talking about metaphors and analogies a lot. To get started, we will first define the basic terminology and explain it with a few examples. In the rest of this section we will then discuss the use of metaphors and analogies in education in general.

1.1.1 Terminology

The terms *analogy* and *metaphor* are often misused in practice. What is possibly even worse, different communities use these terms in slightly different senses. For instance, there is a difference between a metaphor in literature, and a metaphor in cognitive science. We will be more interested in the latter.

Metaphor

Collins English Dictionary [8] gives the following definition:

> Metaphor: a figure of speech containing an implied comparison, in which a word or phrase ordinarily and primarily used of one thing is applied to another (Ex.: "the curtain of night", "all the world's a stage").

Under this definition, almost any figure of speech where we are not using the literal sense can be classified as a metaphor. And actually, in literature this is often the case. But for us such a definition is useless, as it is not related in any way to the way we use metaphors in our thoughts.

We found a much better definition in the area of cognitive linguistics. Below we give our definition, based on the definition by Lakoff and Johnson [15].

> A (conceptual) *metaphor* is a cognitive process that occurs when a subject seeks understanding of one idea (the target domain) in terms of a different, already known idea (the source domain). The subject creates a conceptual mapping between the properties of the source and the target, thereby gaining new understanding about the target.

Note the difference. According to this new definition, the first example given in the dictionary definition does not classify as a metaphor. You cannot really think about nights in terms of what you know about curtains. The second Shakespearean example may probably still be considered a metaphor.

As an example of a good metaphor in computer science, consider the name "flood fill" commonly used for breadth-first search. Anyone who goes to study algorithms can already imagine a flood. And it is actually possible to think about the algorithm in terms of a wave of water that is rushing down all available corridors at the same pace.

Another example of a good metaphor in computer science: topological sort can be nicely introduced via the "getting dressed" metaphor. There are many non-trivial dependencies between various pieces of clothing. Their graph can easily be constructed by students and it is a good example for a test run of the algorithm. The intuitive approach "put on anything you already can" actually translates to a correct and efficient algorithm for topological sorting.

Personification

There is one subtype of metaphors that is worth mentioning separately: the *anthropomorphic metaphor* (*personification*). For a personification, the source domain is a living human being.

Personifications are common in colloquial speech of computer scientists. For instance, you have most probably both heard and used the expressions "the program is running", "the compiler needs to see a semicolon there", or "the client machines periodically ask the server".

Analogy

Once we decided to use the restricted definition of a metaphor, we also needed a new term for the rest of the original set. In accord with cognitive linguistics, we will call them analogies. More precisely:

> An *analogy* is a cognitive process in which a subject transfers information from one particular object to another. The word *analogy* can also be used as a noun describing the similarity between the two particular objects.

A sample analogy: CPU is *like* the brain of the machine *in that* it takes input data, processes it and produces outputs.

By our definition, every metaphor is an analogy, but not vice versa. In an analogy, only some of the properties of the source are transferred to the target; these properties may be explicitly stated as a part of the analogy. In a good metaphor, the target and source should match on all relevant properties, enabling us to infer information about the target in terms of the source. For example, the analogy "the CPU is the brain" would make a poor metaphor—it is impossible to think about the CPU in terms of what we know about our brains. In classroom setting, such poor metaphors may lead to unnecessary confusion.

Abstraction

An *abstraction* is a process by which a subject derives more general, higher level knowledge from patterns observed in multiple particular objects. For example, pseudocode is an abstraction from various imperative programming languages. If a problem solver encounters sufficiently many individual problems for which particular greedy algorithms work, he/she may form an abstraction: a strategy of looking for a greedy algorithm.

Abstraction of knowledge is, in some sense, the holy grail of both research and education. A famous example is the case of pecking amongst chicken. Among a group of chicken, they often peck each other to establish order in their group. Pecking is usually asymmetric—if chicken A pecks chicken B, then B does not dare to peck A. But, surprisingly, it was observed that it is not necessarily transitive. For example, there may be a trio of chicken such that A pecks B, B pecks C, and C pecks A.

Still, it turned out that even in chicken groups with such non-transitive behavior it was still possible to find a "linear pecking order"—i.e., to arrange all chicken into a sequence in which each chicken pecked the next one. Empirical evidence about this was first published as research in biology. However, the surprise ended with Landau's proof [16] that this is, in fact, a mathematical certainty that has nothing to do with chicken. In the language of graph theory, Landau has shown that every tournament graph contains a Hamiltonian path.

1.1.2 Metaphor as a Teaching Tool

On a very simple and basic level (and with a necessary apology to all researchers in the area of education), the process of learning can be described as follows: learning is trying to understand stuff we do not know using stuff we already do know. From this point of view, good metaphors and analogies are an ideal teaching tool: the students quickly get the big picture, and their efforts to grasp the new concept are supported by their past experience.

For instance, in English the literal meaning of the word "hot" denotes high temperature. However, the same word is also used to describe spicy foods. This is an example of an analogy being actually adopted into everyday usage. And it is a good analogy: after a child hears the sentence "the pizza with chilli peppers is really hot", it will have a reasonably good idea what would happen after eating a mouthful.

Note that we did not classify this as a metaphor. Why? The only property that actually gets transferred is the taste. Without any other experience, the child in the example may wait for ten minutes "to let the pizza cool down" and then try eating it anyway.

Overview of Past Research

As we stated above, metaphors and analogies are an important tool in learning unfamiliar concepts. Many authors agree that they can be used not only to communicate the relevant properties of the new concept, but also to facilitate developing new conceptual structures—new mental models and new abstractions. For a more in-depth treatment of analogies and mental models we recommend [10]. Below we present an overview of past research that is specifically related to the topic of this book: metaphors and their use in education, in particular education of computer science.

As the first topic in our overview, we would like to address a particularly interesting question: at which level of education is it suitable to use metaphors? We are convinced that the answer is simple—at all levels. There is no inherent reason to stop using metaphors once we reach a particular difficulty level. And the same is true at the opposite end of the spectrum: it is possible to use metaphors in education from a very early age.

Of course, in computer science most of the documented uses of metaphors can currently be found in materials for undergraduate courses that teach the basics of the actual computer science. For instance, it is well documented that the notion of a variable is often confusing for beginners in programming. Statements like "`x = x+1`" tend to make no sense, and sequences of instructions such as "`a=1; b=a+1; a=9`" tend to be misunderstood (with students claiming that `b=10` at the end).

A particularly successful way of dealing with these issues is to introduce the concept of a variable using a suitable metaphor [3]. Probably, the most famous one is the simple "labeled box" metaphor—a variable is box used to store a value, and you can refer to the value by the label on the box. However, this is not the only choice. Another usable metaphor (for variables that store numbers) is the "traveller"—the variable is a traveler that sequentially visits different numeric locations.

Many nice metaphors and analogies useful for teaching computer science to young kids (even pre-school and elementary school students) can be found in [1, 22]. For pre-university and undergraduate levels, we recommend [11] as an excellent introduction to computer science. The author uses some metaphors: For instance, in one of the early chapters programming is explained in terms of baking a cake.

In the existing literature on computer science (and on teaching computer science), there is an abundance of metaphors for simple concepts and a seeming lack of ones that target more involved concepts. This disparity is easily explained: In order to devise a good metaphor, the educator has to have a sufficient amount of experience in the particular area. And the more difficult concepts we consider, the fewer experts there are. To support our claim that metaphors can be used at all levels of education,

1.1 Metaphors in Education

we present two metaphors that are being successfully used to teach concepts at a graduate level of university education of computer science.

The first of these cases are the "potential energy" and the "piggy bank" metaphor [6], both being used when analyzing the amortized time complexity of data structures. It is easier for the students to reason about this involved concept in terms of known basic concepts from physics and/or accounting.

The second notable case is the "Ali Baba's cave" metaphor [17] used to illustrate the basic mechanisms behind a zero-knowledge proof: the prover must convince the verifier about his/her knowledge via an interactive protocol, but at the same time a casual onlooker must not gain any information about the secret knowledge.

Another question we should address in this overview is the question whether using metaphors actually helps. In [2], the authors provide empirical evidence for the claim that good explanatory analogies are useful in teaching programming. According to the authors, the good analogies are characterized by clarity and a systematic approach. The research had shown that such analogies positively influence all facets of education, including program comprehension, program composition, and time needed for the given tasks.

In his doctoral thesis, Woollard [21] focuses on the role of metaphors in the teaching of computing. In the thesis, the author attempts to divide metaphor usage into several distinct forms. In particular, the most prominent form is the narrative theme, where an object, a function or a system is described "in the clothes" of a different, more familiar object. Other uses of metaphors include algorithms, models, diagrams, and role play. The author also analyzes various aspects of metaphor usage, and concludes by outlining future research proposed to determine the effectiveness and efficiency of particular metaphoric strategies.

In [20], the author presents visual metaphors that represent simple programming concepts (such as data types, variables, files, etc.). It is shown that these metaphors accelerate the students' learning process and improve their comprehension of programs as structured objects.

According to [5], computer science in particular is a special area. The way how computer science metaphors work does not fit neatly into prevailing general theories of metaphor. Computer science metaphors provide a conceptual framework in which to situate constantly emerging new ontologies in computational environments. The author argues that these metaphors have a unique role in learning, design, and scientific analysis. Computer science metaphors trade on both pre-existing and emerging similarities between computational and traditional domains.

Caveats When Using Metaphors

One point to remember when using metaphors and analogies in education is that they are very sensitive to cross-cultural aspects. Keränen in [14] discusses various cross-cultural aspects of using metaphors in computer science education. Duncker in [9] gives a detailed account of one particular case where native Maori struggle to understand the library metaphor which is foreign to their experience and way of living. Duncker also briefly discusses other culturally biased metaphor, such as the North American mailbox with a flag set when incoming mail is present.

Another issue, especially of interest in technical fields, is the gender bias of some metaphors. For instance, "car analogies" are commonly used in engineering courses and textbooks such as [4]. By our experience, these tend to be less suitable for female students.

Yet another source of difficulties with metaphors is the fact that the target never matches the source perfectly. The inherent danger is that a subject will use the metaphor to infer incorrect conclusions about the target. Such conclusions are very dangerous, as the subject usually possesses a high confidence in these conclusions. Even when using good metaphors, this danger must be foreseen and mitigated by the educator.

This aspect is often overlooked in education. As we show in later sections, many teachers and textbook authors will gladly use a metaphor (such as "imagine the queue data structure as a checkout line in a store" in [13] and many others) without considering its flaws. Some of those metaphors turn out to be fundamentally flawed—they lead the students to incorrect conclusions about important aspects of the new concept. In such cases, we strongly recommend to present the new concept using a clearly outlined analogy instead of a metaphor.

Finally, we would like to point out that while metaphors are a good tool, they should never be used *exclusively*. Even when a metaphor seemed to work perfectly, it may be the case that the subject will later encounter some more subtle cases in which thinking about a concept *solely* in terms of a metaphor can sometimes lead to incorrect conclusions. Spolsky [18] quotes several prominent cases where simple metaphors fail in unexpected ways. One prominent example is the two-dimensional array (a table).

The two-dimensional array is a basic data structure available in many programming languages. This is a mathematical abstraction that allows us to organize our two-dimensional data in a logical way. For most basic purposes, it is possible to think about the data structure in the same way as about a grid of cells on paper.

However, note that the names "two-dimensional array" and "table" are just metaphors, not accurate descriptions of reality. They have nothing in common with the way the data is actually stored in physical memory. And this difference does sometimes come into play.

For instance, imagine that a programmer needs to iterate over all elements stored in such an array. When thinking in terms of metaphor, the programmer will come to the obvious (but wrong) conclusion that iterating in row major order should be the same as doing it in column major order. In the mathematical abstraction, this is obviously equivalent—we process each element once. But in computer programs this is not the case: row major order is usually a order of magnitude faster due to caching. (It is faster to process data in the order in which it is stored in physical memory.)

This example highlights the need to complement a metaphor-based explanation using other techniques. Spolsky concludes that becoming a proficient programmer is getting harder and harder, even though we have higher and higher level programming tools with better and better abstractions.

1.2 Metaphors and Computers

Before we start discussing the more involved metaphors in computer science, we present some of our observations on simpler examples. In particular, in this section we discuss various basic metaphors commonly associated with computer usage. We use them to illustrate the dangers of a flawed metaphor.

As soon as you turn on your computer, you are bound to run into metaphors. In fact, almost all common user interfaces use metaphors to make the interface as accessible to beginners as possible. To start with, you have a *desktop*, where you can organize your *files* into *folders*. Those are not real files and real folders in the original sense of those words, and there is no desk inside your computer—but your prior experience with the physical objects allows a new user to infer the roles of these new, virtual objects.

The *folder* metaphor is of a particular interest. In the first hierarchical filesystems, this object was usually called a *directory* or a *catalog*. Those are metaphors, too, but with a different purpose. The *directory* metaphor highlights the properties important to an engineer—the corresponding record in the filesystem contains a list of files (and possibly also subdirectories) that are present at the corresponding place in the hierarchy. On the other hand, the name *folder* (made popular by early releases of Windows) is directed at the end user, making it clear that a folder is an object that can contain other objects, mostly files. For the end user, the name *directory* can, in fact, be confusing, as they will not perceive a connection between actual directories and those on their computer.

Some of the metaphors clearly work. A *menu* lists the choices you have, and allows you to read them and pick the one you want. A *bookmark* works just like an actual bookmark, bringing you quickly to a place that you marked before. A *check box* is a box you can check, just like on a form you fill in with a pen.

It is important to note that a metaphor is neither inherently good, nor inherently bad. For example, metaphors can sometimes become outdated. They may have been good in the past, but that is not the case any more. For instance, consider *radio buttons*—the controls on a form that allow you to select exactly one of the presented options. Our enquiry among our students (of ages 20 and less) showed that many of them just learned, at some point in time, that "those are called radio buttons" and that was it. For them, this was not a metaphor at all—because they were missing the original concept it refers to.

The name *radio buttons* comes from old radios that had a bunch of actual physical push-down buttons that worked that way—once you pushed one of them, it caused the other ones to pop out, so just one button could be pushed down at any time. For the generation that coined this metaphor, the metaphor worked. But for a generation where the radio is an application on a touchscreen cellphone, the metaphor does not work at all.

This illustrates a more general principle we already mentioned before: that all metaphors are, in fact, culturally biased. Many of the seemingly basic concepts of one culture are absent or different in other cultures. Hence, a metaphor that is nice and appropriate for one culture could be useless in other parts of the world.

Another thing to keep in mind: Not all cases when an old word is given a new meaning are necessarily metaphors. As a prominent example, the name *mouse* for the pointing device was coined simply from its appearance. The user was never expected to think about computer mice in terms of the real ones.

In the previous section, we briefly mentioned flawed metaphors and the dangers associated with their usage. In daily computer usage we can find many metaphors that are confusing (and sometimes even dangerously so) for many users. For years, in MacOS (prior to OSX) a common way to eject a CD-ROM drive was to drag its icon to Trash. The "desktop" and "trash" metaphors were so strong that many users were anxious about losing their data when "dragging the CD to trash" [12]. Stephenson [19] gives another example of a flawed metaphor, so common that proficient computer users do not even realize that it is a metaphor: the words "document" and "save". When we document something in the real world, we make fixed, permanent, immutable records of it—but computer documents are mutable. And if we "save" something in real life, we protect it from harm. But every time you hit "save" on your computer, you annihilate the previous version of the document—the one you supposedly "saved" before.[1]

We can also find flawed metaphors in all areas of computer science. For instance, in Sect. 4.1 we discuss the *queue* data structure in detail. As a part of this discussion we show that the popular "checkout line" metaphor is fundamentally flawed and should be avoided—or at least used with care, and only as an analogy.

Another frequently used but flawed metaphor is the Russian nesting dolls—"Matryoshkas". These nesting dolls are commonly used as metaphors for two distinct concepts: for recursion and for nesting of data structures [7]. In our experience, both usages turned out to be problematic. With recursion, the main caveat is that each of the dolls always contains at most one smaller doll—this sometimes caused the false impression that a function can only make one recursive call. With nested data structures, the main misconception that arose from the metaphor was that the structures have to be of the same type.

1.3 How to Read The Main Chapters

In the main chapters of this book we present a selection of topics from various areas of computer science. For a better readability, we tried to present each topic using more-or-less the same structure. This structure and our intentions behind choosing it are explained below.

---

[1] For this particular metaphor things are taking a turn for the better—some of the modern tools already "auto-save" your documents and actually store the entire edit history, so that each state of the document is actually saved, in the original sense of the word.

Overview. We will usually start by giving a brief overview of the topic. Here, we define the problem and provide the minimal background necessary to follow the next sections.

In some cases, the "traditional" presentations of the problem and/or its solutions contain various types of pitfalls. Where applicable, we also include those in this section.

Metaphor. This is the core part in which we develop our metaphor and show how it can be used to teach the topic.

Analysis. In this section we address the topic from the scientific point of view. If necessary, we fill in the details necessary to get from the metaphor to a complete solution of the problem. Afterwards, we provide a more involved analysis of the topic, discuss the best solutions known, and cite references to related research publications.

Experience. Whenever we can, we also try to include a separate section with helpful didactic notes for the teachers and/or an account of our experience when using the currently discussed metaphor in classroom setting.

Exercises. We always conclude a topic by presenting several related exercises. Good exercises serve two purposes. For the student they provide an opportunity to apply the newly acquired knowledge. For the teacher, the students' performance when solving the exercises can serve as an indicator whether they actually have a deep understanding of the currently discussed topic.

References

1. Bell, T., Fellows, M.R., Witten, I.: Computer Science Unplugged... Off-Line Activities and Games for All Ages. http://csunplugged.com/ (1998). Cited 8 Dec 2012
2. Chee, Y.S.: Applying Gentner's theory of analogy to the teaching of computer programming. Int. J. Man Mach. Stud. **38**(3), 347–368 (1993)
3. Chiu, M.M.: Metaphorical Reasoning: origins, uses, development and interactions in mathematics. Educ. J. **28**(1), 13–46 (2000)
4. Cohoon, J.P., Davidson, J.W.: C++ Program Design: An Intro to Programming and Object-Oriented Design, 3rd edn. McGraw-Hill, Boston (2002)
5. Colburn, T.R., Shute, G.M.: Metaphor in computer science. J. Appl. Logic. **6**(4), 526–533 (2008)
6. Cormen, T.H., Leiserson, C.E., Rivest, R.L., Stein, C.: Introduction to Algorithms, 3rd edn. MIT Press, Cambridge (2009)
7. Danzig, N.: Introduction to computer science—C++. http://www.danzig.us/cpp/ (2009). Cited 8 Dec 2012
8. Dictionary.com: Collins English Dictionary—Complete & Unabridged, 10th edn. Harper-Collins Publishers, New York (2012). Cited 8 Dec 2012
9. Duncker, E.: Cross-Cultural usability of the library metaphor. In: Proceedings of the 2nd ACM/IEEE Joint Conference on Digital Libraries (JCDL 2002), pp. 223–230. ACM (2002)
10. Gentner, D., Stevens, A.L. (ed.): Mental Models. Lawrence Erlbaum Association, Hillsdale (1983)
11. Hromkovič, J.: Algorithmic Adventures. Springer, Berlin (2009)

12. Hübscher-Younger, T.: Understanding algorithms through shared metaphors. In: Proceedings of the CHI 2000 Conference on Human Factors in Computing Systems (CHI 2000), pp. 83–84. ACM (2000)
13. Keogh, J.E., Davidson, K.: Data Structures Demystified. McGraw-Hill, New York (2004)
14. Keränen, J.: Using metaphors in computer science education—cross cultural aspects. Tech. Rep., CS Department, Univ. of Joensuu (2005)
15. Lakoff, G., Johnson, M.: Metaphors We Live By. University of Chicago press, Chicago (2003)
16. Landau, H.G.: On dominance relations and the structure of animal societies III. The condition for a score structure. Bull. Math. Biophys. **15**(2), 143–148 (1953)
17. Quisquater, J.J. et al.: How to explain zero-knowledge protocols to your children. In: Proceedings of the 9th Annual International Cryptology Conference (CRYPTO 1989), pp. 628–631. Springer (1989)
18. Spolsky, J.: The Law of Leaky Abstractions. Joel on Software. http://www.joelonsoftware.com/articles/LeakyAbstractions.html (2002). Cited 8 Dec 2012
19. Stephenson, N.: In the Beginning was the Command Line. Harper Perennial, New York (1999)
20. Waguespack, L.J., Jr.: Visual metaphors for teaching programming concepts. In: Proceedings of the 20th Technical Symposium on Computer Science Education (SIGCSE 1989), pp. 141–145. ACM (1989)
21. Woollard, W.J.: The rôle of metaphor in the teaching of computing; towards a taxonomy of pedagogic content knowledge. Ph.D. Thesis, London (2004)
22. Yim, K., Garcia, D.D., Ahn, S.: Computer science illustrated: engaging visual aids for computer science education. In: Proceedings of the 41th Technical Symposium on Computer Science Education (SIGCSE 2010), pp. 465–469. ACM (2010)

Chapter 2
Graph Algorithms

2.1 Single-Source Shortest Paths in Graphs

2.1.1 Overview

Possibly, the most useful graph algorithm is the computation of a shortest path from a given vertex A to a given vertex B. In addition to obvious applications such as "what is the fastest way of traveling from city A to city B", this algorithmic question appears in many unexpected places.

For instance, imagine the well-known Rubik's cube puzzle: a $3 \times 3 \times 3$ cube with colored faces. Solving this puzzle can be modeled as a graph problem: each of the approximately 4.3×10^{19} different configurations can be seen as one of the vertices of a huge graph. Edges in this graph represent valid moves. When trying to solve a given cube, we are looking for a path from the vertex that corresponds to its current configuration to the vertex that represents the solved cube. The shortest path would then be the solution that requires the fewest moves.

Of course, we may reverse the direction of search. If we start at the vertex that represents the solved cube, we may be interested in the shortest paths to all other vertices—especially the most distant ones. Those represent, in some sense, the configurations that are hardest to solve.

The question about the distance from the starting vertex to any of these "hardest to solve" vertices can easily be stated in words: "What is the smallest number of moves sufficient to solve *any* Rubik's cube?".

It seems impossible to explore a graph of this size—even storing it explicitly would require close to a zettabyte (10^{21} bytes) of space. Still, in 2010 Rokicki et al. [15] used a combination of mathematics and clever algorithms to reduce the search space and then spent 35 CPU-years (a few weeks of computation on a cluster of machines) searching for shortest paths in the rest of the graph. The result? The hardest configurations require precisely 20 moves.

There are plenty of variations to the shortest path problem. In the previous example, the graph was directed (but symmetric, as for each move we have a move that undoes it) and all edges had unit lengths.

In the most general version, the graph can be arbitrary (i.e., directed, possibly with duplicate edges and self-loops) and the edge lengths can be any real numbers—even negative length might emerge in certain situations. An edge of a negative length in a graph is not a problem by itself. However, once a graph contains a cycle of negative length (i.e., the total length of the edges in the cycle is negative), the search for a shortest distance between certain vertices becomes meaningless. More precisely, whenever there are vertices u and v such that there is a path from u to v that is incident to a negative cycle, we can find a walk from u to v with an arbitrarily small length—we just have to extend the path by traversing the negative cycle as many times as we want to. Each walk around the cycle decreases the total length of the walk by a constant.

In practice, instances with negative edge lengths are quite rare. In most practical cases, all edge lengths are positive. Therefore, in this section, we will assume that all our edges have positive lengths.[1]

Single-source shortest paths in a graph

Instance: A (directed or undirected) graph with positive edge lengths and a source vertex s.

Problem: Find the shortest paths from s to all vertices of the graph (Fig. 2.1).

The famous Dijkstra's algorithm is one of the canonical algorithms that can solve this problem. Since there are many variations to this algorithm, in the following paragraph we shall give a brief reference description. For details regarding correctness and efficiency of Dijkstra's algorithm, see [6] or any other standard textbook.

Fig. 2.1 An undirected graph with positive edge lengths, and the shortest path from A to B

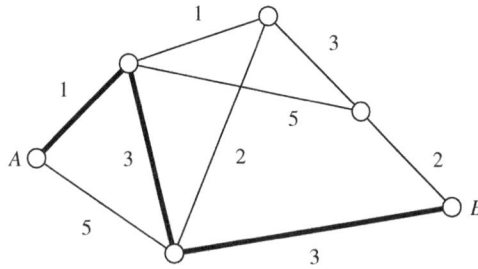

---

[1] Actually, we could just assume that the lengths are non-negative, the algorithm works also in the cases when some edge lengths are zero. Still, this is just a trivial detail and for the sake of our metaphor we prefer strictly positive edge lengths.

2.1 Single-Source Shortest Paths in Graphs

The input for Dijkstra's algorithm is a graph with positive edge lengths and one marked starting vertex s. The algorithm computes the lengths of shortest paths from s to each of the other vertices of the graph. The algorithm proceeds as follows: Let the *best known distance* of vertex v be the smallest known distance from the vertex s to v. These values are stored in the array D; the value $D[v]$ is updated whenever a smaller distance to v is discovered. Initially, all $D[v]$ are infinite except for $D[s] = 0$. Each vertex v is in one of two states: finished (meaning that $D[v]$ is final) or active. Initially, all vertices are active. The algorithm then repeats the following steps:

1. Let u be an active vertex with the smallest distance value, i.e., the active vertex that is closest to the vertex s.
2. The vertex u becomes finished and thus $D[u]$ is final.
3. For each edge uv: update $D[v]$ if the path from s via u to v is shorter than $D[v]$. (The length of this new path is $D[u]$ plus the length of uv.)
4. If there are still some active vertices left, continue with step 1.

2.1.2 Metaphor

There is a well-known metaphor related to the shortest path problem: the balls-and-strings metaphor (Fig. 2.2). Each vertex of the graph corresponds to a small ball. The balls are connected by strings with lengths proportional to edge lengths. For instance, one textbook mentioning this metaphor is [7], but the metaphor is not actively used in algorithm presentation.

One well-known exposition of the shortest path using the balls-and-strings model looks as follows: To find the shortest path between s and t, one just grabs the corresponding two balls and tries to pull them apart. This will be stopped when some strings between the balls become stretched. Then the distance between the two balls corresponds to the length of the shortest path between s and t and the stretched strings then form the shortest path.

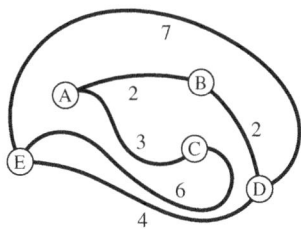

Fig. 2.2 A balls-and-strings model of a graph: the balls correspond to vertices, the strings correspond to edges, their lengths are proportional to the length of edges in the graph

We will now show how to augment this metaphor by adding the aspect of gravity. This will add multiple benefits:

- The above process only works for a single-source, single-target version of the problem. Our metaphor addresses the single-source, all-targets version.
- Our metaphor illuminates how all the single-source shortest paths look like.
- Most importantly, using our new metaphor we can easily describe a physical process that directly corresponds to the inner workings of Dijkstra's algorithm. The entire algorithm can then easily be explained (and reasoned about) in terms of the balls-and-strings metaphor. We note that similar insights can then lead to new algorithm design, as in [13].

Balls, Strings, and Gravity

The change needed to introduce our version of the metaphor is very simple: Instead of pulling a pair of balls apart, we take the balls-and-strings model of the graph and we let it hang by the ball corresponding to s. For the graph shown in Fig. 2.2, the result is shown in Fig. 2.3.

In this very simple way we already managed to generalize the metaphor to single-source all-targets shortest paths: gravity is pulling all the balls away from s at the same time. Once the model is stable, for each of the other balls there has to be some sequence of stretched strings between s and that particular ball—otherwise gravity would pull the ball even further down. It is easily seen that the set of strings with no slack corresponds precisely to the edges of all shortest paths from s to other vertices in our graph.

Fig. 2.3 A balls-and-strings model of a graph from Fig. 2.2 that is hanging by the ball A. The stretched strings correspond to the edges present in the shortest paths from A to all other vertices (In the picture, the stretched strings are slightly non-vertical in order to make it easier to read. In practice, those strings would be vertical)

2.1 Single-Source Shortest Paths in Graphs

Therefore, to find the shortest distances from s to all other vertices, it is sufficient to determine how the model looks like in its stable state, hanging by the ball that corresponds to s. If we actually had the physical model, we could simply hang it on a wall and look at it to find all the answers.

Adding New Strings

Instead of just hanging the entire model on the wall at once, we shall later describe a simple physical process that *builds* the model in small incremental steps. The final result will be the same as above: we will have the balls-and-strings model of the entire graph hanging on the wall by the vertex s. But before describing the actual building process, we shall first make two simple observations.

Suppose that we already have some balls-and-strings model hanging on the wall. Imagine that you are adding a new piece of string between two of its balls. Obviously, this can never pull a ball *further downwards*—at any moment, each of the balls is as low as it can, due to gravity. There are only two possible cases. If the string is long enough, both balls it connects remain where they were. If the string is short, it will pull the deeper ball upwards from its current depth.

The other observation we shall need later: the addition of a new string *does not influence anything above the two balls it connects*. Once again, this is obvious thanks to our intuition from physics.

Building the Model Incrementally

Below we explain the steps of building the model. For the graph shown in Fig. 2.2, all the steps of the building process are shown as subfigures in Fig. 2.4. The reader is encouraged to follow the explanation along with the corresponding subfigures.

Initialization. We start by fixating the ball that represents the starting vertex on the top of a vertical wall. (In our example, the source vertex s corresponds to the ball A. This ball gets attached to the top of the wall in Fig. 2.4a.)

Incremental building. We then process the model from top to bottom. Whenever we encounter a ball U, we add all the strings that should be connected to ball U and are not connected to this ball yet. The opposite end of each of those strings also gets attached to the corresponding ball. (For instance, in Fig. 2.4b, when ball A was processed, we added new strings to balls B and C.) In this way, new balls may become attached to the model, and already present balls may get more strings.

Properties of the Incremental Algorithm

Completeness. We should show that the above process will actually build the entire model. (More precisely, in cases when the original model was disconnected, we will build the entire component that contains the starting ball.)

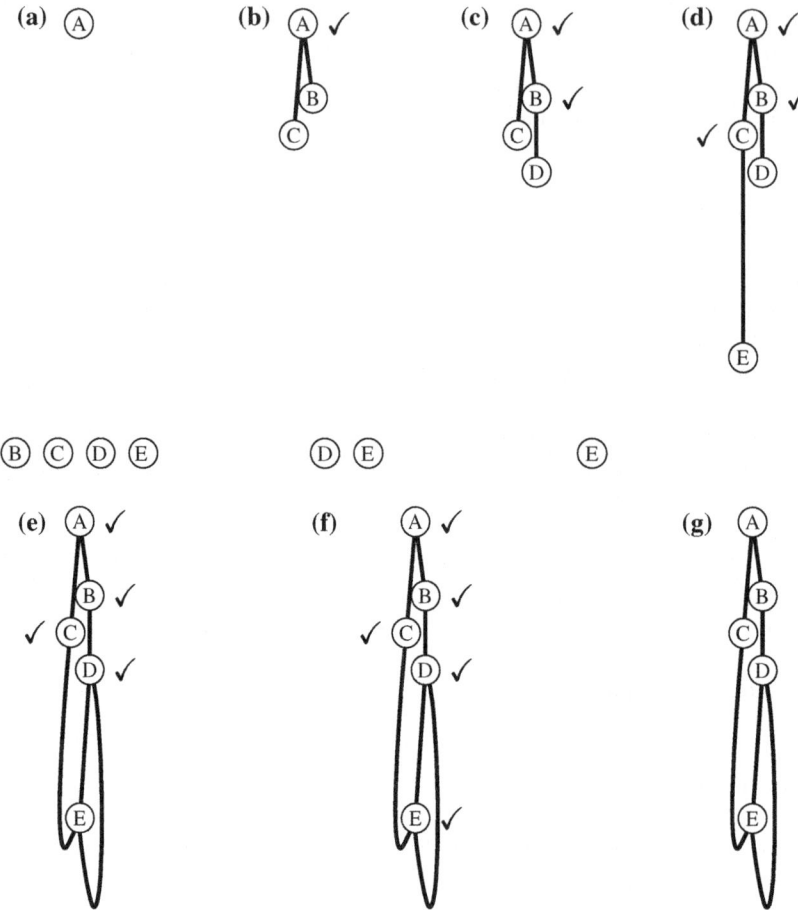

Fig. 2.4 The individual steps of building the graph from Fig. 2.2. The most important step to notice is shown in Fig. 2.4e: when D is finalized and a new string between D and E is added, the ball E is lifted upwards—the distance between E and A decreases.
(**a**) Start: ball A is fixed on the wall, other balls lie on the ground;
(**b**) Processing A: A becomes done, balls B and C get attached to A;
(**c**) Processing B: B becomes done, ball D gets attached to B;
(**d**) Processing C: C becomes done, ball E gets attached to C;
(**e**) Processing D: D becomes done, E is *lifted up* by the new string to D;
(**f**) Processing E: E becomes done, no new edges are added;
(**g**) The final outcome: the entire balls-and-strings model on the wall

All we need for that is to show that each ball that gets added to the model will also be processed at some later point in time. Once we know that, it follows that the model contains the original ball, all of its neighbors, all of their neighbors, etc.

Suppose that we are processing some particular ball U. It may happen that during this step we connect some new ball V to the model we are building. (In our example,

2.1 Single-Source Shortest Paths in Graphs

this happens in Fig. 2.4b–d.) At this moment, the new ball is below the ball that is currently being processed, so we expect it to be processed at some point in the future.

There is only one theoretical possibility for the ball V *not* to be processed—while we are processing some other ball W, the ball V would have to get above W. But that is clearly impossible: when processing W, all the strings we are adding have both ends at the level of W or below W. And by adding such strings, we can clearly never lift V above W.

Finalization. During the incremental building, once we process a ball and declare it final, the ball will remain in its current place. This follows directly from the above observations: after some ball U is declared done, all the next balls and strings will be added below U. These additions cannot influence U in any way—U cannot go further downwards, and the added balls and strings clearly cannot lift U upwards.

Final shape. The final shape of the balls-and-strings model is determined only by the string lengths. And as we know that (for connected graphs) the incremental process will build the entire model, at the end the model will have exactly the same shape as when we hanged it on the wall all at once. In other words, once we finish building the model, the current depths of all balls represent the shortest path lengths, as desired.

(In our example, the final state shown in Fig. 2.4g is exactly the same as the model shown in Fig. 2.3.)

Connection to Dijkstra's Algorithm

Finally, we shall show that the incremental process of building the model in most aspects directly corresponds to Dijkstra's algorithm (as described on p. 13).

- The fixation of the initial ball on the top of the wall corresponds to the initialization of Dijkstra's algorithm: the distance for the starting vertex is set to zero.
- The processing of balls from top to bottom precisely matches the way how Dijkstra's algorithm processes the vertices by increasing distance from s (the step 1).
- The addition of a string from the current ball U to a different, not-yet-processed ball V represents the consideration of edge uv in the step 3 of Dijkstra's algorithm.

The parallels go deeper. In particular, consider the following ones:

- The first discovery of a new vertex corresponds to the moment when the ball is connected to the hanging part of the model for the first time.
- When adding a piece of string that will hang with some slack, the depths of balls do not change. But when adding a short piece of string, we have to pull the deeper ball upwards. This is precisely the moment when one of the shortest distances gets updated in step 3 of the algorithm.

  Also note that in such cases we always only lift a single ball—as it has not been processed yet, there are no other balls hanging below it.

- The addition of new strings can never pull a ball further downward. As we add new edges to the graph, the lengths of shortest paths can only decrease.
- Once a ball is processed, all changes to the model happen below the ball, therefore its depth never changes. This is the exact reasoning we need to prove correctness of Dijkstra's algorithm: once a vertex v is declared finished, its value $D[v]$ is final, because it cannot be improved via any of the unfinished vertices.

2.1.3 Analysis

There are many different algorithms that can be used to find a shortest path from a single source vertex to all other vertices in a graph. The time and space complexities of these algorithms depend on properties of the given graph: we will use n to denote its number of vertices and m for the number of edges.

If the graph is unweighted, the fastest algorithm is a simple breadth-first search which takes time $O(n + m)$.

In the most common case where the edge lengths are arbitrary non-negative reals, Dijkstra's algorithm is the best choice. In its simplest implementation with adjacency matrix used for edge storage, its time complexity is $O(n^2)$. If the graph is rather sparse and an adjacency list is used to store edges, its time complexity is decreased to $O(m \log n)$. With a Fibonacci heap this can be pushed further down to $O(m + n \log n)$.

Once we allow edges with negative lengths, Dijkstra's algorithm stops being usable. (Some implementations of Dijkstra's algorithm will give incorrect outputs for such instances, others will require exponential time, and they may even run forever in the presence of negative-length cycles.)

In the presence of negative edges, one can use the Bellman-Ford algorithm both to check for the presence of negative-length cycles and to compute lengths of single-source shortest paths if such a cycle is not reachable from the source vertex. The algorithm is also based on a dynamic programming technique and it can also be implemented efficiently in distributed environments. The time complexity of this algorithm is $O(nm)$.

All of the algorithms above can also be used for graph with directed edges. Additionally, if a directed graph does not contain any cycles, we can use a dynamic programming algorithm similar to topological sort to compute the single-source shortest paths in the graph even if the edge lengths are arbitrary (possibly negative). The time complexity of this approach is $O(n + m)$, which is better than Dijkstra's algorithm for such cases.

See [6] for a more thorough overview and analysis of path search graph algorithms.

2.1.4 Experience

In our experience, the metaphor is very clear and accessible to students from very different backgrounds and cultures. The only slightly negative aspect are the occasional comments from students on the impossibility to construct the physical model with 100% accuracy. But even the complaining students usually understood it as primarily a thought experiment.

When using the metaphor instead of the traditional textbook definition of the algorithm, we observed that our students grasped the topic faster and easily gained a reasonably deep understanding of the algorithm. This could be tested by posing suitable questions, such as the ones presented below as exercises. Here is one other question we often use: "Why exactly can the algorithm fail in the presence of negative edges? Can you design an instance *without* negative edges that will cause the algorithm to fail?"

As can be seen in the solutions to the exercises, the metaphor gives new non-trivial insights into the algorithm. For instance, once we realize that in the hanging balls-and-strings model some strings have slack while others do not, we have gained insight into how the shortest paths look like.

2.1.5 Exercises

2.1. Given is a graph, its vertex s, and its edge uv. Design an algorithm that will check whether there is a vertex t such that some shortest path from s to t uses the edge uv.

2.2. Given is a graph, its vertex s, and its edge uv. Design an algorithm that will check whether there is a vertex t such that removing the edge uv will increase the length of the shortest path from s to t.

2.3. Design an algorithm that will generate all possible shortest paths from the vertex s to the vertex t in a given graph. What is the worst-case time complexity of your algorithm?

2.4. Modify the algorithm from the previous exercise to *count* the shortest paths from s to t instead of generating them. The time complexity of the modified algorithm has to be polynomial.

2.5. Assume that we already computed the shortest distances from a given vertex s to each other vertex in a graph. Now suppose that the length of one edge of the graph suddenly decreased. Design an algorithm that will check whether some of the shortest distances changed or not.

2.6. Continuing the previous exercise, design an algorithm that will compute the new shortest distances from s to all vertices. Your algorithm should *not* process the entire graph, only the part of it that has actually changed.

2.7. Given is a graph with n vertices and m edges. One of the vertices is s. We want to remove as many edges as possible without changing the length of the shortest path from s to any other vertex. Prove that we can always remove some $m - n + 1$ edges, and that this is optimal. Design an algorithm that will find one such set of edges.

2.8. Using the physical balls-and-strings model, scissors and glue, how would you find the length of the *second* shortest path from s to t? Does the approach you suggested translate into a polynomial algorithm for graphs?

2.2 Longest Paths in Trees

2.2.1 Overview

Trees are one of the most fundamental structures in computer science.

Tree-like data structures are among the most effective ones for representing ordered sets of elements. Problems related to trees were among the first ones investigated in computer science.

One such problem is the minimum spanning tree problem—given a graph with weighted edges, we want to use some of its edges to build the cheapest possible tree that connects all vertices. This problem was first investigated by Czech mathematicians when constructing an efficient electricity network for Moravia. The first polynomial algorithm[2] for this problem was discovered by Borůvka in 1926 [2, 3] and later improved by Jarník in 1930 [11]. (Note that the CS Unplugged book [1] has a nice activity related to the minimum spanning tree problem.)

Even if we limit ourselves to simple unweighted trees, we will still encounter many non-trivial problems. For instance, a hard combinatorial question is counting all trees on vertices labeled 1 through n. A famous result by Cayley states that the number of such trees is precisely n^{n-2}. This combinatorial result is closely related to an algorithm "from the Book": the Prüfer encoding of trees [14]. Prüfer gives a simple method that can be used to uniquely encode any such tree into a sequence of $n-2$ integers, each from 1 to n. Cayley's formula [5] follows as a direct consequence. Using modern data structures, Prüfer's algorithm can be implemented in $O(n \log n)$ time.[3]

As the topic of this section suggests, we shall consider another problem related to simple trees.

---

[2] This was three decades before Dijkstra came up with the shortest paths algorithm presented in the previous section. And then it took another decade until Cobham and Edmonds brought our attention to considering polynomial time complexity as a synonym of efficiency.

[3] And even in $O(n \log \log n)$ using the van Emde Boas tree [10].

2.2 Longest Paths in Trees

Fig. 2.5 A small simple tree. One of the longest paths in this tree is highlighted

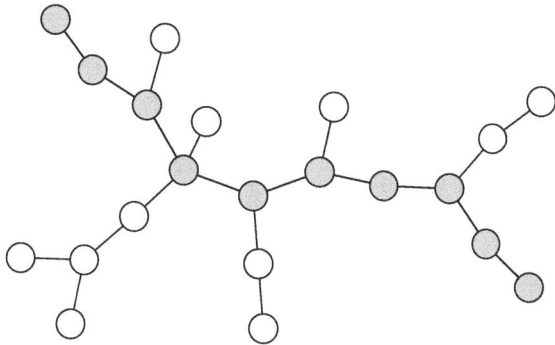

Longest path in a tree
Instance: A simple tree with n vertices and $n - 1$ edges of unit length.
Problem: Find one of the longest simple paths in the tree.

A sample instance and its solution are shown in Fig. 2.5.

In particular, we shall focus on another algorithm that should probably be in the Book. The algorithm is incredibly simple to implement (we just need two traversals of the tree), its time complexity is clearly optimal, but its correctness is far from being obvious. The algorithm looks as follows:

1. Let x be any vertex of the tree.
2. Let y be any vertex that is the farthest from x.
3. Let z be any vertex that is the farthest from y.
4. The path between y and z is a longest path in the tree.

The metaphor presented in the next subsection allows for a simple proof of the above algorithm. Additionally, we shall show a few related results:

- The center of a tree is always formed by one or two vertices.
- Each longest path in a tree contains the center of the tree.
- We can count all longest paths in $O(n)$ (even though there can be $\Theta(n^2)$ of them).

The metaphor is an adapted version of one published in [9]. Our treatment of the metaphor and the additional results is original work.

2.2.2 Metaphor

The balls-and-strings model we used in the previous section works nicely in this setting as well. However, its presentation will be slightly different in order to highlight the inequalities we will need to see in order to find the longest paths in a tree.

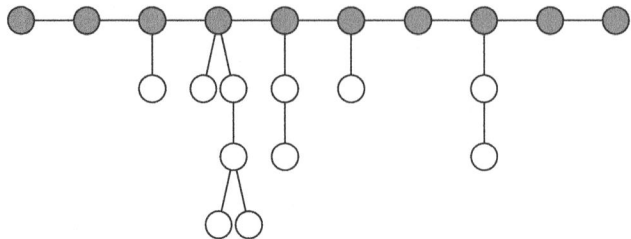

Fig. 2.6 A tree hanging by the ends of one of its longest paths

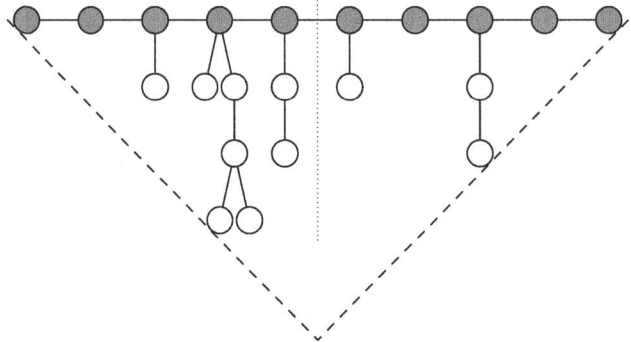

Fig. 2.7 The *dashed lines* encompass the possible locations of all vertices in our tree. The *dotted line* shows the middle of the chosen longest path

Consider an arbitrary simple tree. One such tree is shown in Fig. 2.5. We will assume that we already know a longest path in our tree. We will pick our tree up by the endpoints of the path, and pull them apart as far as possible, so that our path becomes a horizontal chain of vertices. Due to gravity, the rest of the tree now hangs downwards from this chain. For the tree and the path highlighted in Fig. 2.5 we obtain the situation shown in Fig. 2.6.

Of course, the same process can be done with any two vertices. But the fact that we chose the longest path now limits the locations of other vertices to the triangle shown in Fig. 2.7.

Why is that the case? Consider any vertex a in the left half of the tree (e.g., the vertex a in Fig. 2.8). Let a' be the closest vertex of the top path to a. (We will call this vertex the *root* of a.)

Note the two paths highlighted in Fig. 2.8: aa' and la'. The path aa' must be at most as long as la'—the path lr is a longest path, so the path $aa'r$ cannot be longer. Hence the triangular shape—the distance between l and a' limits the depth of the subtree rooted at a'.

In the following text, we will use the term *diagonals* to denote the two nonhorizontal sides of the triangle shown in Figs. 2.7 and 2.8. When talking about *vertices on the diagonals*, we mean all vertices that are as deep as they possibly

2.2 Longest Paths in Trees

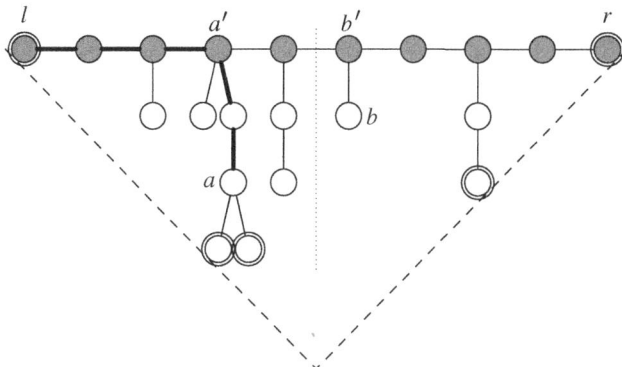

Fig. 2.8 The distance between a and a' must not exceed the distance between l and a'. The same holds for b, b', and r

can, given their root. In Fig. 2.8, the vertices that are on the diagonals are marked by double circles. Note that the set of vertices that lie on the diagonals always includes the vertices l and r.

We can now strengthen the above observation:

> For any vertex x of the tree, all vertices that are the farthest from x lie on the diagonals.

To prove the above claim, we will proceed in two steps: First, we will only consider x located on the chosen longest path, and then we will generalize to all possible locations of x.

For the first step of the proof, consider any vertex x' that lies on the chosen longest path. In our tree (that is hanging by the chosen longest path, as in Fig. 2.8) the path from x' to any other vertex has either the form "some steps left, then some steps down", or the form "some steps right, then some steps down". In the first case, the length of the path is limited by the length of $x'l$—in that many steps we hit the left diagonal and there is nowhere to go after that. In the second case we are limited by the length of $x'r$ in the same way. For instance, in Fig. 2.8 each vertex to the left of a' is at most 3 steps away, and each vertex to the right of a' is at most 6 steps away. Hence, if our x' lies in the left half, the farthest vertices from x' all lie on the right diagonal, and vice versa.[4]

Now consider a vertex x anywhere else in the tree. The path to most other vertices starts by going from x to its root x'. Hence, the farthest of those vertices are those farthest from x', and we already know that those are where we claimed. Now let v be any vertex such that the path xv does not pass through x'. (That is, x and v are in the

---

[4] And if x' happens to be exactly in the middle of the chosen longest path, vertices on both diagonals are in the same distance from x'.

same subtree under x'.) Then the path xv is strictly shorter than the walk $xx'v$, and this walk is obviously at most as long as the longest path in our tree. Hence, such vertex v cannot be the farthest one from x. This concludes the proof.

The correctness of the examined algorithm follows easily. We pick an arbitrary vertex x. We find any vertex y that is the farthest from x. The vertex y is somewhere on one of the diagonals. Now, the distance from y to the opposite end of the chosen longest path is equal to the length of lr, hence there is a longest path that starts at y. Hence by picking any vertex z that is the farthest from y we actually get one of the longest paths, q.e.d.

This part of the proof can also be easily visualized using our metaphor: we can just change the chosen path from lr to ly (or yr) and then to zy (or yz) without changing the length of the chosen path. This is shown in Fig. 2.9.

The center of the tree. Informally, a vertex v belongs to the center of a graph if you can get quickly from v to any other vertex. The formal definition is quite complicated to grasp, as it involves *three* nested quantifiers:

The *center of the graph* is the set of all vertices v that minimize the maximum (over all w) shortest path distance from v to w.

In order to make this definition more approachable, it is advised to introduce the quantifiers one at a time:

- Let $d(v, w)$ be the shortest distance between vertices v and w.
- The *eccentricity* of a vertex v is $\max_w d(v, w)$, i.e., distance to the farthest vertex.

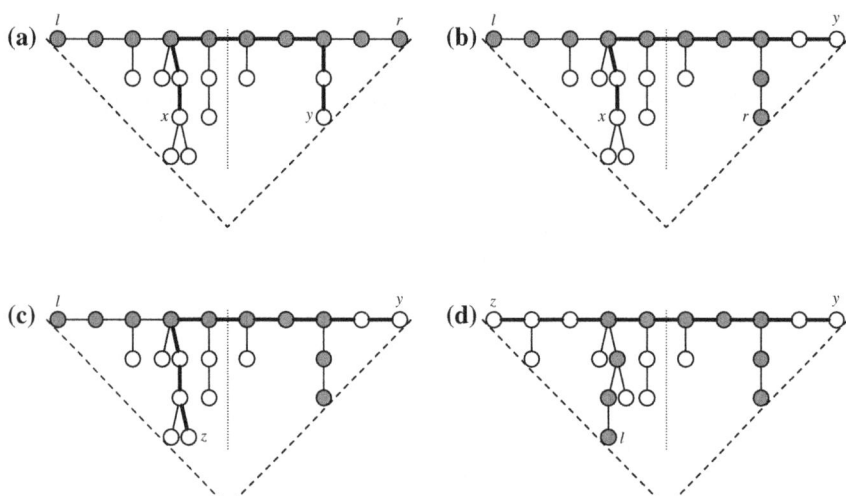

Fig. 2.9 Visualization of a sample run of the longest path algorithm.
(**a**) Steps 1 + 2: pick x; y is farthest from x;
(**b**) After step 2: ly is exactly as long as lr;
(**c**) Step 3: z is farthest from y;
(**d**) Step 4: yz is a longest path

2.2 Longest Paths in Trees

- The center of a graph is the set of vertices with minimal eccentricity.

For our convenience, we add two more definitions:

- The minimal eccentricity value is called the *radius* of the graph.
- The maximal eccentricity value is called the *diameter* of the graph.

Our metaphor makes it very clear how the center of a tree looks like. Let ℓ be the length of the longest path in a tree. (In other words, let ℓ be the diameter of the tree—in trees these two terms coincide.) Immediately, we can explicitly compute the eccentricity for all vertices on the chosen longest path: the eccentricity of the endpoints (l and r) is ℓ, and as we go closer to the center of the path, eccentricity decreases by 1 in each step. The vertices with a smallest eccentricity are the one or two vertices in the middle of the chosen path, and their eccentricity is $\lceil \ell/2 \rceil$.

It is also immediately obvious that the eccentricity of any other vertex is strictly greater than $\lceil \ell/2 \rceil$. We can do even better and compute their exact eccentricities as well—each step downwards in our tree has to increase it by 1, because we are getting farther away from both l and r. And, as we already know, for any vertex x at least one of l and r is among the farthest vertices from x.

The center and the computed eccentricities for our sample tree are shown in Fig. 2.10.

The choice of a particular longest path does not matter—the computed eccentricities will always be the same (they do not depend on our visualization). The classical result follows:

> In a tree, all longest paths contain the center of the tree.

(Another way of showing this result: Consider our tree, hanging by some chosen longest path. Let pq be any longest path in our tree (possibly different from the first

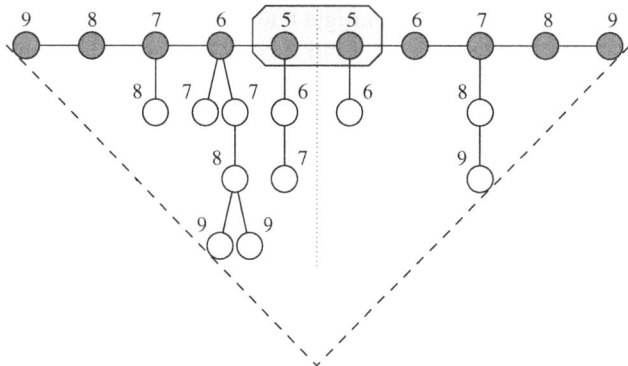

Fig. 2.10 The center of the tree are the two highlighted vertices in the middle of the longest path. The vertex labels are their eccentricities. Also note that the vertices with maximum eccentricity are precisely the ends of all longest paths

one). As q is the farthest from p, it lies on one of the diagonals. The same holds vice versa, hence we conclude the following: Each longest path has one endpoint on the left diagonal, and the other on the right diagonal. It follows that the center of the one chosen longest path is necessarily contained in each other longest path as well—we have to get from the left half to the right one.)

Counting the longest paths. It should now also be obvious how to count all longest paths. As we just stated, each longest path has one endpoint on the left diagonal, and the other on the right diagonal. If the diameter ℓ is odd (that is, if the center contains two vertices), it is enough to multiply their counts—for instance, if there are 3 vertices on the left diagonal and 5 on the right one, each of the $3 \times 5 = 15$ possible pairs determines a different longest path.

Finding the counts of vertices on each diagonal can be done in linear time by running a breadth-first search (or any other graph traversal) starting at both vertices that form the center of the tree. Alternately, the above algorithm to compute the eccentricities of all vertices can easily be modified to count these vertices.

For ℓ even the situation requires a more careful analysis, but at this point this analysis can be easily carried out by the students, and we list it below as one of the exercises.

2.2.3 Analysis

The first discovery of the tricky algorithm presented in the Metaphor section is unknown. Some authors [8, 12] attribute it to Wennmacker, others [4] to Dijkstra.

As noted in [12], with an actual physical balls-and-strings model, locating the longest path can easily be done using gravity. In our even simpler version: pick up any ball with your left hand, then catch the lowest ball by your right hand, release the ball in your left hand, and finally catch the currently lowest ball by your left hand, and you are done. We opted *not* to use this exposition—while it is stunningly beautiful, it does not offer sufficient insight into *why* the algorithm works.

The algorithm is clearly optimal—its time complexity is linear in the number of vertices. Additionally, its implementation is clean and concise, all we need is a tree traversal procedure. By the nature of a tree, any traversal works, including, but not limited to, breadth-first search and depth-first search.

Below we give a shorter but less visual and less intuitive proof of the algorithm. That is, we will show that the algorithm always finds a longest path in the given tree.

Let xy be the path found in the first graph traversal and let lr be any longest path in our graph. First, we will show that the paths xy and lr must share at least one vertex. Assume the contrary, i.e., we have the situation shown in Fig. 2.11a. Consider the path from x to l. Let p be the last vertex of this path that lies on xy, and let q be the first vertex that lies on lr. As y is the farthest from x, the path py must be at least as long as the path pqr. Hence, py must be longer than qr. But then the path $lqpy$ is longer lr, which is a contradiction.

2.2 Longest Paths in Trees

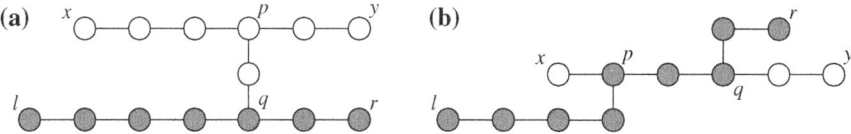

Fig. 2.11 An alternate proof of the two-pass longest path algorithm. Both figures only show the relevant part of the entire tree, there can be other vertices and edges in addition to the ones shown. **a** The case with paths xy and lr disjoint; **b** The case with paths xy and lr intersecting

Now, we know that xy and lr share at least one vertex, as shown in Fig. 2.11b. We will now show that there has to be some longest path that ends at y.

Let p be the first and q the last vertex of xy that is also on lr. If necessary, swap the labels of l and r so that $lpqr$ becomes a simple path. Now, as lr is a longest path, we get that qr is at least as long as qy. On the other hand, as y is the farthest from x, it is necessary that the opposite inequality holds as well: qy must be at least as long as qr. Hence qy and qr are equally long. But then $lpqy$ is a valid longest path.

To conclude the proof, just note that as there is a longest path ending at y, the second iteration of our algorithm will indeed produce one such longest path yz.

There are also other asymptotically optimal algorithms (but with a less simple implementation) that solve this problem. The one we encountered in most textbooks is an algorithm based on dynamic programming: Root the tree at any vertex. For each vertex, compute the depth of its subtree. Then, for each vertex compute the length of the longest path that has its highest point in that vertex. In general, this length is two plus the sum of depths of two deepest subtrees rooted at sons of the given vertex.

2.2.4 Experience

The two-pass algorithm to find the longest path in a tree is so simple that many students were astonished that it works: "How can something so trivial solve such an involved problem?"

Any proof of the algorithm has to deal with the structure of all longest paths in a tree somehow. The presented way of using the balls-and-strings model serves primarily to limit the number of possibilities to examine. It also allows us to visualize the non-trivial constraints that arise once we know a single optimal path. This definitely helps, but the analysis of the algorithm is still far from being trivial, and some students do get lost in the details.

For this problem we hope that the metaphor presented above is *not* the best one possible. In other words, we hope that a different point of view can make the analysis of the algorithm even simpler.

2.2.5 Exercises

2.9. All of the figures in this section contain a tree with an odd diameter. Draw corresponding figures for some tree with an even diameter.

2.10. If the diameter of the tree is even (and hence the center has a single vertex), there can be some vertices that lie on both diagonals at the same time—in other words, vertices that are located in the bottom corner of the bounding triangle. Carefully verify that our proof of the longest paths algorithm still works.

2.11. Give a detailed algorithm that counts all longest paths in a tree with an even diameter.

2.12. Suppose that we have a weighted tree. That is, each edge of the tree has some positive length. Does the presented longest path algorithm still work correctly?

2.13. Suppose that we have a general undirected graph with unit-length edges. The longest path algorithm (using breadth-first search twice) does not necessarily work in this case. Can you find a counterexample?

2.14. As in the previous question, the input instances are now all general undirected graphs with unit-length edges. Suppose that we improve the longest path algorithm as follows:

1. Let x be any vertex of the graph.
2. Let y be any vertex that is the farthest from x.
3. Let z be any vertex that is the farthest from y.
4. If the length of yz is greater than the length of xy: let $x, y = y, z$, goto 3.
5. Output the length of the path between y and z.

Assume that each execution of step 2 and 3 involves a single breadth-first search from the starting vertex. What is the worst-case time complexity of this algorithm in terms of the number n of vertices of the input graph?

2.15. Does the algorithm given in the previous question always find the diameter of the input graph? Prove it, or find a counterexample.

2.16. Is there a known polynomial-time algorithm to find the diameter of a general graph with unit-length edges? Or is this a known NP-complete problem?[5]

References

1. Bell, T., Fellows, M.R., Witten, I.: Computer Science Unplugged: Off-Line Activities and Games for All Ages. http://csunplugged.com/ (1998). Cited 8 Dec 2012
2. Borůvka, O.: O jistém problému minimálním (About a certain minimal problem). Práce mor. přírodověd. spol. **3**, 37–58 (1926) (in Czech)

---

[5] Or, more precisely, is the decision version of this problem NP-complete?

References

3. Borůvka, O.: Příspěvek k řešení otázky ekonomické stavby elektrovodních sítí (Contribution to the solution of a problem of economical construction of electrical networks). Elektronický obzor **15**, 153–154 (1926) (in Czech)
4. Bulterman, R.W., van der Sommen, F.W., Zwaan, G., Verhoeff, T., van Gasteren, A.J.M., Feijen, W.H.J.: On computing a longest path in a tree. Inf. Process. Lett. **81**, 93–96 (2002)
5. Cayley, A.: A theorem on trees. Q. J. Math. **23**, 376–378 (1889)
6. Cormen, T.H., Leiserson, C.E., Rivest, R.L., Stein, C.: Introduction to Algorithms, 3rd edn. MIT Press, Cambridge (2009)
7. Dasgupta, S., Papadimitriou, C., Vazirani, U.: Algorithms. McGraw-Hill, New York (2006)
8. Dewdney, A.K.: Computer recreations. Sci. Am. **252**, 18–29 (1985)
9. Diks, K., Idziaszek, T., Łacky, J., Radoszewski, J.: Looking for a Challenge? The Ultimate Problem Set from the University of Warsaw Programming Competitions. Lotos Poligrafia Sp. z o.o. (2012)
10. van Emde Boas, P.: Preserving order in a forest in less than logarithmic time. In: Proceedings of the 16th Annual Symposium on Foundations of Computer Science (FOCS 1975), pp. 75–84. IEEE Computer Society (1975)
11. Jarník, V.: O jistém problému minimálním (About a certain minimal problem). Práce Mor. Přírodověd. Spol. **6**, 57–63 (1930) (in Czech)
12. Moscovich, I.: The Monty Hall Problem & Other Puzzles. Sterling Publishing, New York (2004)
13. Narváez, P., Siu, K.Y., Tzeng, H.Y.: New dynamic SPT algorithm based on a ball-and-string model. IEEE/ACM Trans. Netw. **9**(6), 706–718 (2001)
14. Prüfer, H.: Neuer Beweis eines Satzes ber Permutationen (A new proof of a theorem about permutations). Arch. der Math. u. Phys. **27**, 742–744 (1918) (in German)
15. Rokicki, T., Kociemba, H., Davidson, M., Dethridge, J.: God's Number Is 20. http://www.cube20.org/ (2010). Cited 8 Dec 2012

Chapter 3
Computational Geometry

3.1 Shortest Path with Obstacles

3.1.1 Overview

Our first problem from the area of computational geometry is a famous optimization problem: finding the shortest path that avoids a given set of obstacles.

> **2D shortest path with obstacles**
> **Instance:** A collection of non-overlapping obstacles, and two points (A and B) that lie outside of all obstacles.
> **Problem:** Find one shortest curve from the source point A to the destination point B, given the constraint that the curve must not contain any inner points of the obstacles.

Two-dimensional shortest paths represent a class of very important problems in computational geometry, with applications in motion planning (see Chap. 13 in [2]). In particular, we will be interested in the case where all the given obstacles are polygons. We will use n to denote the total number of vertices in all the polygons. A sample instance of the problem is depicted in Fig. 3.1.

The main issue we have to address is the number of possibilities. As this is a continuous problem, the number of possible paths is infinite. (More precisely, not even countable.) In order to find a shortest path, it is necessary to identify a finite set of *candidate paths* that certainly contains at least one shortest path. Once this reduction is done, the set of candidates can be algorithmically processed.

We will show a metaphor that nicely explains this reduction and leads directly to a polynomial-time solution for the case of polygonal obstacles in the two-dimensional plane.

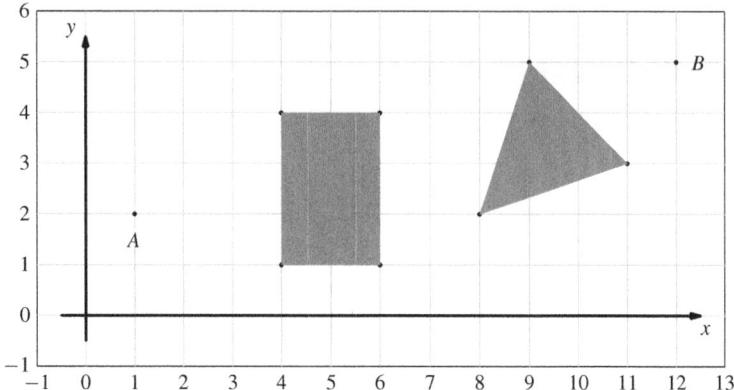

Fig. 3.1 An instance of the 2D shortest path with *obstacles*. The *obstacles* are *polygons* and their total number of *vertices* is $n = 7$

3.1.2 Metaphor

Directly by their nature, shortest path problems are the ideal playground for the "rubber band" metaphor. A rubber band is a physical object that has one intuitive property—it always tries to contract. If we hold its ends in place, the body of the rubber band will become stretched, in order to be as short as possible. (In all our uses of this metaphor, we consider an idealized version of the rubber band that can contract to a point if allowed to.)

Our use of this metaphor aims to illuminate the reduction of a seemingly infinite number of possible paths into a small set of candidates. To do so, one has to determine some properties of the shortest path. In particular, the following statement will be useful: The shortest path can always be divided into segments such that each segment is either a straight line segment, or it copies the boundary of an obstacle.

The intuition behind the proof of this statement is nicely illuminated by a clever use of the rubber band metaphor:

Fix one end of a rubber band in a starting point A. Keep the other end in your hand and walk toward point B using any path you like. As you walk, the rubber band will stretch in your hand as it will try to contract back to the origin point A. Once you reach the point B, it will be at most as long as your actual path was. In other words: *if* your actual path was optimal, *then* the rubber band must exactly follow its trajectory. An example is shown in Fig. 3.2.

Note that the contracting rubber band does not necessarily produce the globally shortest path. The final length of the rubber band depends on the path taken. (Formally, the rubber band will form the shortest of all paths that are topologically equivalent to the path you took.) Refer to Fig. 3.3 for an example.

For polygonal obstacles we now easily obtain a stronger corollary:

3.1 Shortest Path with Obstacles

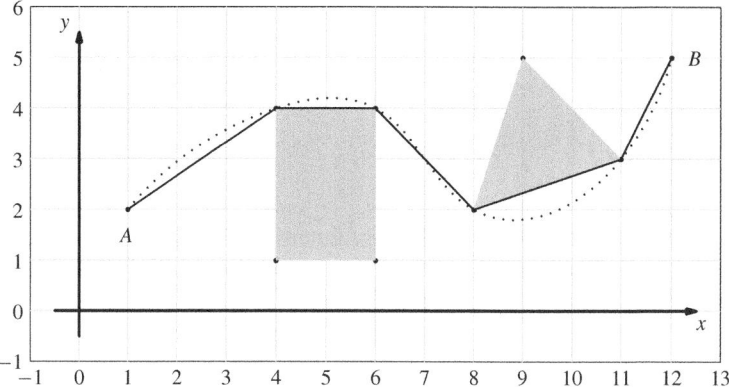

Fig. 3.2 If one walks with the *rubber band* from point *A* to *B* along the *dotted curve*, the *rubber band* stretches into the *solid line*

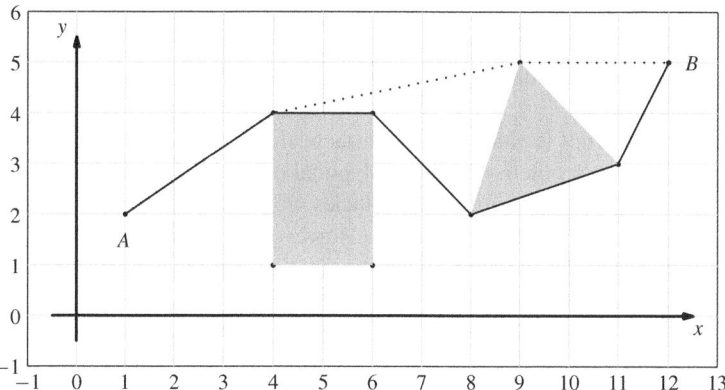

Fig. 3.3 The *dotted path* is the shortest path between points *A* and *B*. However, if one walks from *A* to *B* roughly along the *solid line* (for example as in Fig. 3.2), the *rubber band* will contract to form the *solid line*. Its length will be longer than optimal

---

The shortest path in a two-dimensional plane with polygonal obstacles is always a polyline that starts in A, ends in B, and all the intermediate vertices are some of the vertices of obstacles.

Again, the proof is easily illustrated by the rubber band metaphor. As the polygon sides are line segments, from the above statement it follows that the rubber band is a polyline. Now, the rubber band will never change direction if not forced by the environment. Intuitively, it is easy to convince ourselves that the changes of direction will only occur in vertices of the obstacles, never on their sides. (An actual proof is given in the next section.)

3.1.3 Analysis

As stated in the previous section, in the two-dimensional shortest path with obstacles problem the shortest path can always be divided into segments such that each segment is either a straight line segment, or it copies the boundary of an obstacle.

A formal proof of this statement can be based on local optimizations. Obviously, for the optimal path there can be no possible local optimizations. Hence, the shape of the optimal path must not allow any local optimizations. Consider any section of the optimal path such that none of its points lie on an obstacle boundary. We will prove that this section has to be a straight line segment. Assume the contrary. Pick any point where the optimal path is not a straight line. Then there has to be some small-enough neighborhood of this point that does not contain any obstacles. Within this neighborhood we can improve the path by replacing it by a line segment. This is a contradiction with the optimality of the path. Hence, the optimal path must indeed consist only of sections that copy the boundary of an obstacle and sections that are straight line segments.

Note that this proof is valid for any obstacle shapes, not only for polygons. The issue that makes the problem with non-polygonal obstacles harder is the localization of segment endpoints. For polygonal obstacles we already know that it is sufficient to consider the vertices of the obstacles. However, if we allow general curves as obstacle boundaries, the endpoints of the straight line segments can be located almost anywhere on the curves—see Fig. 3.4 for an example.

For polygons, a formal proof of the stronger corollary also follows from local optimizations: if you have a polyline that has a vertex on the side of a polygonal obstacle, you can always shorten the polyline by shifting the vertex of the polyline toward one of the vertices of the obstacle. An example is shown in Fig. 3.5.

Now we can define the *visibility graph*. The vertices of this graph are the points A and B, and also all the vertices of all the obstacles. Two vertices are connected by an edge if and only if the segment connecting them avoids all obstacles. See Fig. 3.6 for an example of a visibility graph.

Using the above observations, we can now change the original, continuous version of the shortest path problem into a finite discrete one: finding the shortest path in the visibility graph. This solution can easily be implemented with a polynomial time complexity. The graph version of the shortest path problem can be solved in $O(n^2)$ using Dijkstra's algorithm. The entire visibility graph can be constructed in $O(n^3)$ time using brute force—for each pair of vertices, test for intersections between the line segment that connects them and each obstacle.

There are also faster ways of constructing the visibility graph. For instance, a polar sweep for each vertex accomplishes this in $O(n^2 \log n)$ (see e.g., Chap. 15 in [2]).

3.1 Shortest Path with Obstacles

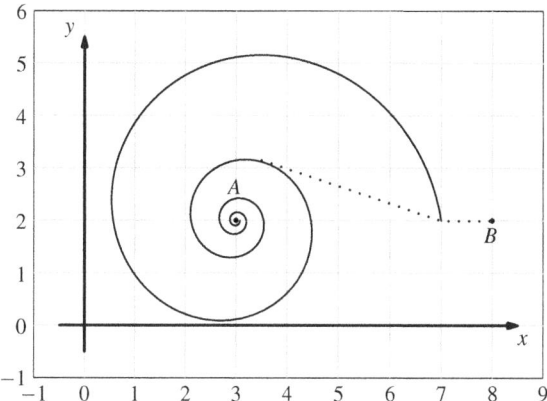

Fig. 3.4 An *obstacle* having the form of a *spiral* wall. The shortest path goes from A to the *inner endpoint* of the *spiral*. Then it follows the boundary of the *spiral* for a while. Finally, it leaves the *spiral* along a *tangent* that touches the *outer endpoint* of the *spiral*. From that point it concludes by a *straight segment* to B. The last two parts are shown as the *dotted line* in the figure

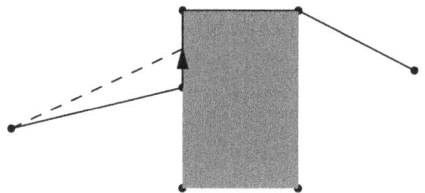

Fig. 3.5 The optimal path will never have a *vertex* on the *side* of an *obstacle*. Thanks to the *triangle inequality*, any such path can be locally improved. Moving the *vertex* of the *polyline* in the direction of the *arrow* shortens the path

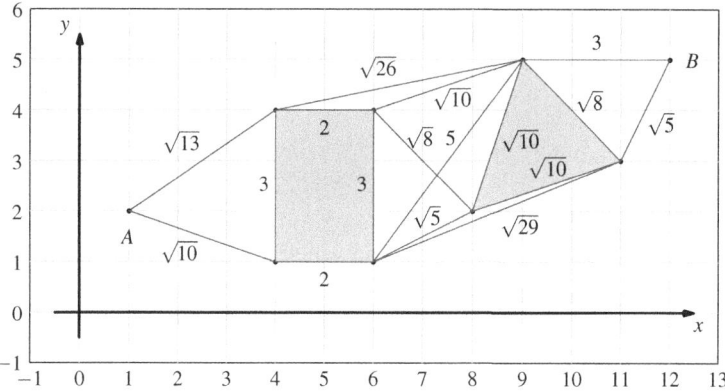

Fig. 3.6 The *visibility graph* for the instance from Fig. 3.1. It contains 9 *vertices* and 17 *edges*. Edge lengths are Euclidean distances

An even clever plane sweep yields an algorithm that constructs the visibility graph in $O(n^2)$, see [1, 7, 17].

During the many years of research, by more insightful approaches the overall time complexity was reduced to the optimal $O(n \log n)$ [10, 11] and for some special cases even to $O(n)$ [4, 8].

Note that while the problem is efficiently solvable in the plane, the three-dimensional version of the shortest paths with obstacles is NP-hard [3]. A short summary of graduate improvements together with references can be found in Sect. 15.4 of [2].

The rubber band metaphor tends to be useful on its own in designing algorithms to solve other, related problems as well [12].

3.1.4 Experience

In addition to just using the metaphor in a direct explanation, it can also be easily used as a kinesthetic activity. We suggest doing it in the following way: Chairs and tables can be used to form a small maze of obstacles. Fix the points A and B and have the students actually try walking along different paths, while holding to one end of a rubber band. The student will experience the stretching and they will observe the behavior of the band as described above.

Actually, it is simpler to do the activity using a plain piece of string in place of a rubber band. (The ones in real life are not infinitely elastic.)

One possible way of doing the activity with a string (or a rope): One end of the string will be anchored in the point A. The student is asked to hold the other end and walk to B. Upon reaching B, the student then pulls on the string to remove any slack. We can then make a mark on the string that corresponds to the currently measured distance. (Alternately, we can prepare a string with distance marks and then use them to read off the distance between A and B.)

The teacher can also initiate a game to find the shortest path from A to B— have multiple students try different paths, measure each of them and evaluate the experiment.

3.1.5 Exercises

3.1. Consider a variation of our problem in which the obstacles are disjoint circles. Can our algorithm be modified to work in this case?

And how about the case when both circular and polygonal obstacles are allowed? (For simplicity, you may still assume that all obstacles are pairwise disjoint.)

3.2. Rubber bands work in three dimensions as well. Explain why the metaphor does not yield an efficient algorithm in the case of polyhedral obstacles in three dimensions.

3.1 Shortest Path with Obstacles

(Hint: Sometimes we need to cross over an edge of a polyhedron.)

3.3. If you found the previous exercise too hard, here is a simpler one that still illustrates the difficulty of finding shortest paths in three dimensions:

Given is an orthogonal box and two points on its surface. Find the shortest path that connects them without entering the box.

(Hint: Imagine that you mark the shortest path onto the surface of the box, and then unfold the visited sides of the box into a plane. How must the marked path look like after the unfolding?)

3.4. The length of the shortest path between A and B is *numerically stable*—minor changes in the input correspond to minor changes in the output. However, the actual shape of the shortest path is not. Can you find an instance for which a minor change of an obstacle yields a significant change of the shortest path?

3.5. Let us get back to the two-dimensional case with polygonal obstacles. As a warmup, prove that when the point B is moved by x in some direction, the distance between A and B increases by at most x.

Now assume that you were given an instance of our problem, and that you already used the algorithm explained above to construct the visibility graph and to find the shortest paths from A to all vertices of the visibility graph. Afterwards, you are given new information: the point B should be moved by at most x. For simplicity, we will assume that there is no obstacle within the distance x of B, so all directions of movement are available.

Design an algorithm that will decide whether it is possible to move B in such a way that the shortest distance between A and B will actually increase by x.

What is the optimal time complexity in which this problem can be solved?

3.2 Distance Between Line Segments

3.2.1 Overview

Determining the shortest distance between two line segments is one of the basic problems in computational geometry that is used as a primitive in many more involved algorithms.

> **Distance between line segments**
> **Instance:** Two line segments (in two or more dimensions).
> **Problem:** Find the shortest straight-line distance between them.

Obviously, as the input size is constant, there should be a way to determine the distance in constant time. There most certainly is such a way. Its main idea is rather

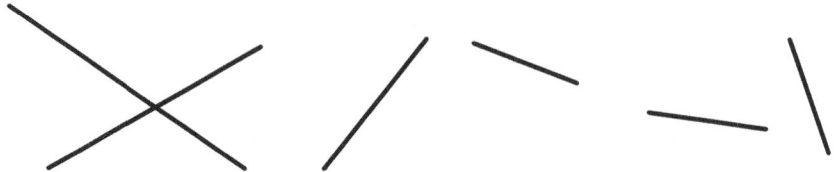

Fig. 3.7 Three distinct cases occur in *two dimensions*. On the *left*, the *two line segments* intersect and their distance is zero. In the *middle* the shortest distance is between a *pair of endpoints*. On the *right*, the shortest distance is between *one endpoint* and its perpendicular projection to the *other line segment*

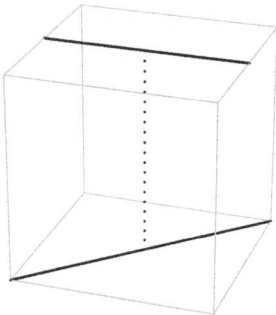

Fig. 3.8 One of the more involved cases in *three dimensions* which students often miss. For a better perception the *two line segments* (*black lines*) are located on a *cube* surface. The shortest connecting *straight line* between the *two line segments* is represented by the *dotted line*

straightforward—consider all the possible cases, find the one that applies, and compute the distance using the corresponding formula.

But then, why is this problem worthy of our attention? Because of the casework. As is the case with many algorithms in computational geometry, the number of various special cases is rather large. Already in two dimensions many students struggle to get all of them right. The three possible cases are shown in Fig. 3.7.

In three dimensions the number of cases increases, and missing some of the cases is rather easy. One of the cases that is notoriously easy to miss is shown in Fig. 3.8. In this tricky case the shortest path that connects the two segments does not contain any of their endpoints—it is a line segment perpendicular to both given segments.

The aim of our metaphor is to show a concise way that leads to identifying all possible cases with ease, even in multiple dimensions.

3.2.2 Metaphor

We will again employ the rubber band intuition. Imagine each line segment as a railroad track. Place one tiny railroad car on each segment and connect them by a

3.2 Distance Between Line Segments

Fig. 3.9 The *rubber band* metaphor: *segments* are imagined as *railroad tracks*, *tiny railroad cars* are connected by a *rubber band* and placed on the *tracks*

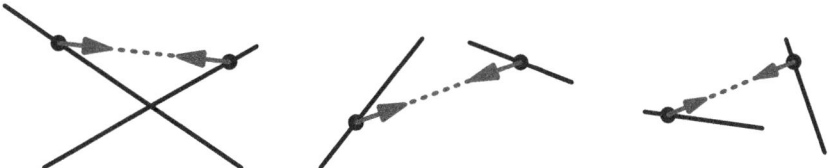

Fig. 3.10 The forces affecting the *railroad cars* in different situations

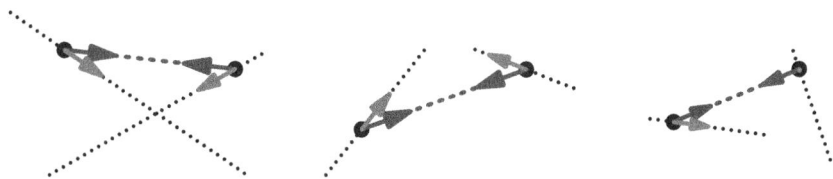

Fig. 3.11 The *orthogonal projections* of forces onto the segments. Note that for the *railroad car* on the *very right* the relevant component of the force is zero. This *car* is not being pulled in any direction at the moment

rubber band (or a bungee cord, if you please), as shown in Fig. 3.9. (Obviously, the railroad cars should be considered to be points in a more precise analysis.)

Regardless of the current position of the two railroad cars, the stretched rubber band always tries to contract, thereby pulling the cars closer to each other. And it is easy to see that this will usually happen—in most configurations the cars will move in one direction or the other. To realize this, it is sufficient to consider the cars one at a time and to visualize how the force applied by the rubber band affects them. This is shown in Fig. 3.10.

As we know from physics, each of the forces can be decomposed into two parts: one orthogonal to the track and one in the direction of the track. We can ignore the first part. It is the second component that affects the movement of the railroad car along the tracks. This decomposition is shown in Fig. 3.11.

Note that both components of the force can be determined as orthogonal projections of the force vector. In the first case we project onto the normal vector of the segment, in the second case we project the original force onto the actual segment.

Fig. 3.12 The shortest distances for our three sample *two-dimensional* instances

Now, whenever the relevant component of the force that affects a car is nonzero, the car gets pulled in one of the two possible directions, and the rubber band contracts—i.e., the cars are moving closer to each other.

What can we tell about the optimal configuration, when the two cars are as close to each other as possible? The answer is simple—the cars are not moving any more!

(Note that what we just discovered is an implication: *if* the location of the cars is optimal, *then* neither of them moves. At this moment, we do not actually know or care whether there are other, non-optimal configurations in which both cars are stationary. See the first two exercises for more on this topic.)

Given what we already know, the algorithm to compute the shortest distance between the two segments starts to take shape. We just have to identify all situations in which neither of the two railroad cars is forced to move, and pick the best one.

Consider any single railroad car, and the rubber band leaving it. In which cases is the car not being forced to move? We can list them easily:

1. The rubber band is already at zero length. Both railroad cars share the same location (and therefore their segments necessarily share that point).
2. The rubber band is orthogonal to the railroad car's segment, so the relevant component of the force is zero.
3. The railroad car is already stuck at an endpoint of the segment. The rubber band tries to pull it further, but there are no more tracks for the car.

In the optimal configuration each of the two cars is in one of the three above cases. By removing the impossible cases and using symmetry, we are left with four possible cases for the pair of cars:

1. The line segments share a common point, and both cars are at one such point.
2. Each car is at an endpoint of its segment.
3. One car is at an endpoint of its segment. The rubber band is perpendicular to the other segment.
4. The rubber band is orthogonal to both segments.

The first three cases are shown in Fig. 3.12. Note that in the above analysis we never needed to specify the number of dimensions. The cases we just derived apply in two, three, and even more than three dimensions.

A simple implementation of this algorithm actually considers all cases that match the above description (regardless of whether they are optimal or not) and picks the one in which the two cars are closest to each other. By our analysis, that case is guaranteed to be optimal.

3.2 Distance Between Line Segments

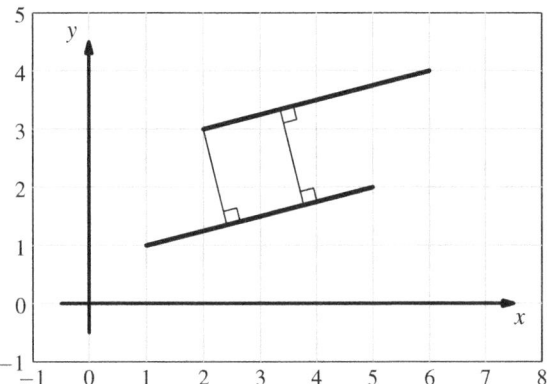

Fig. 3.13 In *two dimensions* case 4 (distance *orthogonal* to *both segments*) is always covered by case 3 (*one endpoint of the shortest distance is a segment endpoint*)

The resulting algorithm for both two and three dimensions, in pseudocode:

1. Check whether the line segments share a common point. If they do, their distance is zero.
2. For each endpoint P of the first segment, for each endpoint Q of the second segment: Compute the distance between P and Q.
3. For each endpoint P of each segment: Compute its orthogonal projection P' onto the line of the other segment. Check whether the projection lies on the other segment. If yes, compute the distance between P and P'.
4. Check whether there is a segment s orthogonal to both our segments (and with each endpoint of s lying on one of our segments). If there is one, compute its length.

Finally, note the following simplifications: We can completely omit the fourth step if we are in two dimensions. This is because that situation only occur if the segments are parallel and the orthogonal projection of one of them onto the other is non-empty. However, that case is already covered in the third step, as shown in Fig. 3.13.

For the same reason, we can omit the fourth step in three dimensions if our two segments are parallel. This makes the calculation simpler, as it gives us a guarantee that there is exactly one direction orthogonal to both of our segments.

3.2.3 Analysis

This Analysis section will be different from most of the other ones. We will not discuss scientific research related to this problem—there are no specific publications that address such a simple problem. Also, the time complexity of an optimal solution is obvious—it has to work in constant time.

Instead, we will use this section to go over the technical parts of our algorithm and show how they can be implemented nicely. We need the following geometric primitives:

- Compute the Euclidean distance between two points.
- Find the orthogonal projection of a point onto a line.
- Check whether a point lies on a segment.
- (In three dimensions:) Check whether two segments are parallel.
- (In three dimensions:) Find a line orthogonal to two given non-parallel segments.

Computing the distance between two points is easy in any-dimensional space.

Figure 3.14 shows the second geometric primitive we need: computing the orthogonal projection of a point (Q_0 in the figure) onto a line ($P_0 P_1$ in the figure). Again, regardless of the number of dimensions, orthogonal projections are easy to compute using a suitable dot product (scalar product) of vectors. Recall that the dot product of two unit vectors is the cosine of their angle. Hence, the orthogonal projection of a point onto a line can be computed as follows: Let $\mathbf{u} = Q_0 - P_0$ and let $\mathbf{v}$ be the unit vector in the direction of $P_1 - P_0$. Then the distance between P_0 and Q_0' (the projection of Q_0 onto $P_0 P_1$) can be computed using the dot product $s = \mathbf{u} \cdot \mathbf{v}$. Hence Q_0' can be computed as $P_0 + s\mathbf{v}$.

Note that in general Q_0' might *not* be located on the actual line segment $P_0 P_1$. This is why we need the third primitive operation mentioned above: to check whether the computed point Q_0' belongs to the segment $P_0 P_1$. We already know that Q_0' lies on the line $P_0 P_1$, and we know that $Q_0' = P_0 + s\mathbf{v}$. The answer to our question is obvious from this representation: Q_0' belongs to the segment $P_0 P_1$ if and only if $0 \leq s \leq d$, where d is the length of $P_0 P_1$.

This is sufficient to solve the two-dimensional case completely. We know how to check all eight possibilities (four involve pairs of endpoints, the other four involve projecting an endpoint onto the other segment). Once we check them, the shortest of the computed distances is our answer.

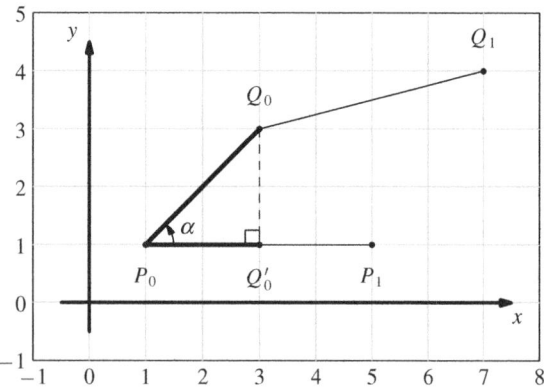

Fig. 3.14 While computing the distance of *segments* $P_0 P_1$ and $Q_0 Q_1$, we need to find the *orthogonal projection* of Q_0 *onto the line* that contains the *segment* $P_0 P_1$

3.2 Distance Between Line Segments

In three dimensions we may still need to consider the last possible situation—the one that applies in Fig. 3.8. We first have to check whether the two segments are parallel. If they are not parallel, the shortest segment that connects them might happen to be perpendicular to both of them.

Let our two segments be P_0P_1 and Q_0Q_1. Let $\mathbf{u}$ and $\mathbf{v}$ be normalized (i.e., unit size) vectors in directions $P_1 - P_0$ and $Q_1 - Q_0$, respectively. We can now represent the line that contains the segment P_0P_1 as the set of points $P_0 + s\mathbf{u}$ for $s \in \mathbb{R}$. Similarly, the other line are the points of the form $Q_0 + t\mathbf{v}$ for $t \in \mathbb{R}$. We can now easily verify whether our lines are parallel by checking whether $|\mathbf{u} \cdot \mathbf{v}| = 1$. If the lines are parallel, the algorithm terminates—as we already know, the parallel case is covered by the cases considered above. Below we address the case when $\mathbf{u}$ and $\mathbf{v}$ are not parallel.

Consider the vector $\mathbf{w} = (Q_0 + t\mathbf{v}) - (P_0 + s\mathbf{u})$. We want to find s and t such that $\mathbf{w}$ becomes perpendicular to both $\mathbf{u}$ and $\mathbf{v}$. Hence, the scalar products $\mathbf{u} \cdot \mathbf{w}$ and $\mathbf{v} \cdot \mathbf{w}$ must both be zero. This gives us a system of two linear equations in s and t. In the non-parallel case this system has a unique solution.

The computed values s and t give us endpoints S and T of the shortest segment that connects the *lines* P_0P_1 and Q_0Q_1. In order to compute the shortest distance between *segments* P_0P_1 and Q_0Q_1, all that remains is to check whether S lies on the segment P_0P_1 and whether T lies on Q_0Q_1. We already know how to do that.

Another way of solving this case is given in [6] by using calculus: The square distance between the points $P_0 + s\mathbf{u}$ and $Q_0 + t\mathbf{v}$ is a quadratic function in both s and t. If the optimal solution is perpendicular to both line segments, we know that each endpoint of the shortest distance segment represents a local minimum on its respective line. Therefore, the first partial derivatives of the square distance according to s and t have to be both zero. From this we can compute s and t.

3.2.4 Experience

While the two-dimensional case is simple enough to be solved without the need for a metaphor, three-dimensional geometry is often tricky and many students have troubles with it (which they usually attribute to their lack of imagination). Our metaphor helps by reducing the number of dimensions—instead of the whole three-dimensional (or even multi-dimensional) problem, we only get two simpler two-dimensional problems.

Another useful tool when presenting this problem was an actual wireframe model of a cube. Such a model is quite easy to construct from some pieces of wire, and it can be very useful as an aid that can be used to visualize the different possibilities for the location of two segments in three dimensions.

Note that we also used such a cube in Fig. 3.8 in order to visualize the three-dimensional case better in a two-dimensional figure.

3.2.5 Exercises

3.6. In the description of the metaphor we only needed the local optimality: *if* the location of the cars is optimal, *then* they do not move. Is there actually an equivalence between those two statements? That is, is it always (regardless of the number of dimensions) true that whenever the two cars do not move, their distance is the shortest one possible?

3.7. Another question related to the previous one: Is there always *exactly one* situation in which neither of the cars is forced to move by the rubber band? If yes, why? If no, is it at least true in two dimensions?

3.8. In two dimensions, what are the possible cases for the shortest distance between two circles?

And between a line segment and a circle? Can you design a complete algorithm to compute their distance?

In either of the two cases above (i.e., circle+circle or circle+segment), can you find a situation when both cars are unaffected by the force of the rubber band, *but* their distance is not optimal?

3.9. In the Analysis section, we described an algorithm that computes the shortest distance between two segments in three dimensions. Does this algorithm also work in *more* than three dimensions? If no, what modifications are needed in order to make it work?

3.3 Winding Number

3.3.1 Overview

In this section and the next one we will be dealing with polygons and polylines.

A closed polyline is a polyline that starts and ends in the same point. Sometimes, we will assign direction to a polyline. Then, a closed polyline can be seen as a sequence of directed line segments such that each segment starts where the previous one ends, and the last segment ends where the first one started.

A polygon is a closed polyline that never touches or intersects itself. The word "polygon" is also used for the entire area enclosed inside such a polyline, including the boundary. (The intended meaning will always be clear from the context.)

We will consider another traditional problem in two-dimensional computational geometry: testing whether a given point is contained in a given polygon.

3.3 Winding Number

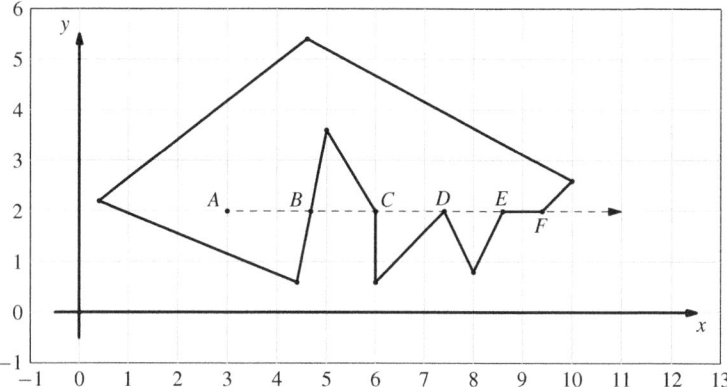

Fig. 3.15 Some of the special cases in the ray casting algorithm: We are checking whether *A* lies *inside* the *polygon*. At *B* the *ray* crosses an *edge* to get out of the *polygon*. At *C* the *ray* hits a *vertex* and reenters the *polygon*. At *D* the *ray* hits another *vertex*, but this time it stays *inside* the *polygon*. Between *E* and *F* the *ray* even goes along *polygon boundary*, before finally leaving the *polygon*

Point in polygon inclusion test
Instance: A point and a polygon (in a two-dimensional plane).
Problem: Check whether the point lies inside, on the boundary, or outside of the polygon.

Checking whether a point lies on the boundary of a polygon can easily be done in linear time: for each side, check whether it contains the point. Below, we shall assume that this test has already been made and that the outcome was negative. That is, the given point is either strictly inside, or strictly outside the polygon.

The canonical solution that distinguishes between those two cases is the "ray casting" algorithm: Pick any ray (i.e., a half-line) starting at the given point. Count the number of times it crosses the polygon boundary. If this number is odd, the point is inside, otherwise it is outside the polygon.

While the idea behind this solution is simple and its correctness obvious, there is a reason we do not really like it: the implementation has too many special cases. Note that we are *not* counting the number of intersections between the ray and the polygon. In some cases, such an intersection counts as crossing the boundary, in others it does not, and in yet other cases there can even be an infinite number of intersections. Some of the tricky cases are shown in Fig. 3.15. In fact, even the original publication of the algorithm [14] contains a bug [9].

A hackish solution to get rid of the tricky special cases is to pick a random direction for the ray. The special cases only occur when the ray hits a vertex of the polygon. And as there are only finitely many directions that hit a vertex of the polygon, with probability 1 the random ray will not hit any of the vertices. However, this approach

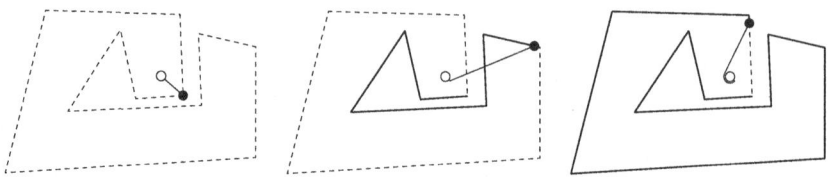

Fig. 3.16 The *girl* (*black circle*) winds the *string around* the *boy* (*white circle*) who stands *inside* the *polygon*. The *left* figure is the situation at the beginning, in the *right* one the *girl* has almost finished her walk

has two disadvantages: In theoretical analysis, a completely correct deterministic solution is preferred. And in practice, picking a random ray will lead to a slower execution than if we were to consider a specific ray (such as a ray in the direction of the x^+ half-axis).

In the Metaphor section we give a simple metaphor that will ultimately lead to a different, less known but more general algorithm: the winding number algorithm. Afterwards, in the Analysis section we discuss its simple implementation. In particular, if all input coordinates are integers, all calculations in our implementation can be done using integers only.

3.3.2 Metaphor

Imagine a boy standing at the given point, and a girl walking once around the boundary of the polygon. Regardless of the actual shape of the polygon, if he happens to stand inside the polygon, the girl will in the end make a complete turn around him. And if he happens to stand outside, she won't.

To visualize this better, let us give the girl a ball of string. She will hand one end of the string to the boy, then she will take her place somewhere on the polygon boundary, and finally she will walk around the polygon boundary once. While she is doing this, she will keep on holding the ball of string and trying to keep the string stretched. The boy will just keep on holding his end of the string and facing her starting point.

It is now easy to convince ourselves about the final outcome. If the boy is outside, the string in the end will look just the same as it did in the beginning. On the other hand, if the boy is inside the polygon, he will end up wrapped in the string. (More precisely, the string will go once around him, and only then toward the girl.)

Two sample runs of the procedure are depicted in Fig. 3.16 (boy stands inside) and in Fig. 3.17 (boy stands outside).

The outcome of the above procedure is intuitively clear but surprisingly hard to prove exactly (as is often the case in topology). If you do not trust your intuition yet, we offer a proof sketch that is easy to visualize. Imagine the polygon as an inflated balloon, forced into its current shape. As we let it go, it will stretch itself and fill in the

3.3 Winding Number

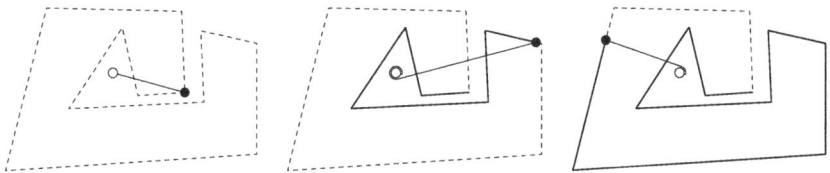

Fig. 3.17 The *girl* does not wind the *string around* the *boy* if he stands *outside*

concave parts, ultimately becoming convex. In this continuous process the outcome of the above procedure never changes, as long as the query point never crosses the polygon boundary.

A formal proof of the above claim can be given using the triangulation of the polygon. For simplicity, assume that the point does not lie on any of the diagonals that form the triangulation. As long as the polygon is not a triangle, its triangulation contains a triangle that shares two edges with the polygon boundary and does not contain the query point. We can replace such a triangle by a straight line segment without changing the outcome of the string winding procedure. After finitely many such changes we are left with a single triangle that either does or does not contain the query point.

Angle sum algorithm. We can now turn the above procedure into our first algorithm. In order to do that, let's first modify the procedure a little bit. The boy will no longer stay still. He is interested in the girl, hence he will be rotating on the spot in such a way that he always keeps facing the girl's current location. Of course, if he does this, he will never get tangled into the string, as the string will always form a straight line between our two actors.

By how much does the boy rotate in this version? We can easily see that the following holds: If the boy stands inside the polygon, all his rotations will add up to exactly 360° (in the same direction in which the girl went around him). And if the boy stands outside, his total rotation has to be 0°.

If that is not immediately obvious, consider the following argument. Imagine that the boy and girl are not doing their actions at the same time. Instead, we will have them perform their actions sequentially. First, we let the girl walk around the entire polygon (and wrap her string around the boy if he happens to stand inside). Only after she finishes will the boy start turning, in the same way as he would in the previous version of the procedure. Clearly, this modification (i.e., postponing the boy's turns until later) does not influence the final outcome of the procedure—it always has to end with a straight string. In other words, what the girl did with the string, the boy then exactly undoes. Hence, his total rotation is 360° if he stands inside and 0° if he stands outside.

The above algorithm can be implemented in linear time: For each side of the polygon, compute the directed angle the boy would turn while the girl walks from one vertex to the other. Sum all those angles. If the result is zero, the boy is outside, if the result is 360° (or −360° if the girl walks in the negative direction), the boy is inside.

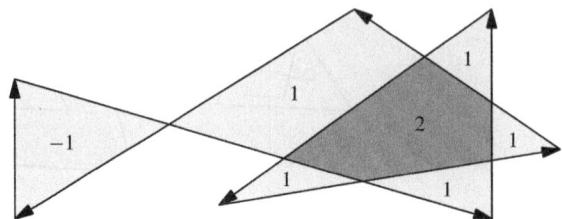

Fig. 3.18 A *closed self-intersecting polyline* and the corresponding *winding numbers*. Note that the *winding number* of a *point* in the *leftmost enclosed area* is *negative*, as the *polyline* winds around that area in *clockwise direction*

This algorithm has a significant advantage: there are zero special cases to consider, and it is obviously numerically stable. On the other hand, computing angles is slower than just intersecting lines, so the implementation has a worse constant factor.

Directed closed polylines and the winding number. Testing whether a point lies inside a polygon is actually just a special case of a more general problem. A polygon is a special case of a closed polyline. If the closed polyline never touches itself, it is obvious that it divides the plane into two parts that can be labeled as the inside and the outside. However, once we allow the polyline to intersect itself, it stops being obvious what "being inside the polyline" actually means.

One possible generalization of the definition is to consider the outcome of our original procedure, in which the girl winds her string around the boy:

The *winding number* of a point with respect to a closed directed polyline that does not contain the point is the number of times the polyline "winds" around the point in the counter-clockwise direction.

Figure 3.18 shows a sample closed polyline and the winding number for points in the enclosed regions of the plane.

Obviously, the angle sum algorithm presented above can be used almost without any modifications to compute the winding number of a given point with respect to a given directed closed polyline. The angle computed at the end is guaranteed to be $w \cdot 360°$, where w is an integer: the desired winding number.

In the following section we first discuss an implementation of this algorithm, and then we combine it with the ray casting algorithm to get a solution that is "the best of both worlds".

3.3.3 Analysis

Without loss of generality we will assume that the boy stands at the coordinates $(0, 0)$. (In the implementation, we can shift the entire scene in order to achieve this.) Now we need to determine the angle by which the boy turns while the girl traverses a segment of the polyline—say from (x_1, y_1) to (x_2, y_2).

3.3 Winding Number

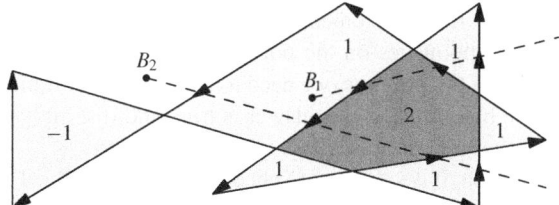

Fig. 3.19 Computing the *winding number* by counting *passes* in *both direction*. The *boy* at B_1 witnesses *two passes counter-clockwise* and *one pass clockwise*. The *boy* at B_2 observes *two passes* in *each direction*. (The *dashed rays* represent the *boys' lines of sight*)

The canonical way of doing this calculation is by using the arctangent function. More specifically, we compute the angle α_1 between the half-axis x^+ and the vector (x_1, y_1), and the angle α_2 between x^+ and (x_2, y_2). The angle by which the boy has turned is now uniquely determined by the fact that it lies in the interval $(-180°, 180°)$ and the fact that it is equal to $\alpha_2 - \alpha_1$ when computing modulo 360°.

Note that $\arctan(y/x)$ has multiple valid return values, so we would need to distinguish among multiple cases according to the signs of y and x. However, most of the modern programming languages offer a function `atan2(y,x)` that uses the signs of y and x to do this for us. Hence, in order to compute the answer for a given polyline with n segments, we need $2n$ calls to `atan2` and $O(n)$ additional simple arithmetic operations.

If we want a faster implementation, we have to get rid of the expensive trigonometric operations. To complete the circle (both metaphorically and literally), we will now introduce an idea similar to the original ray casting algorithm. Instead of having the boy rotate and follow the girl, we will have him looking in a fixed direction and counting the times the girl passes in front of him in each direction. Clearly, regardless of the direction the boy picks, if he counts each clockwise pass as -1 and each counter-clockwise pass as $+1$, his final result has to be the winding number of the point where he stands. This is illustrated in Fig. 3.19.

Of course, we can now simplify the calculations by picking a suitable direction in which the boy will be looking. A natural choice is the half-axis x^+ (i.e., the positive direction of the x axis). At this moment, we have to be careful in order to avoid the type of problems that were shown in Fig. 3.15.

We will say that the girl is in the upper halfplane if she is on or above the x axis, and in the lower halfplane if she is strictly below the x axis. This asymmetry helps us get rid of all special cases: Whenever the girl crosses from the upper halfplane to the lower halfplane through the x^+ half-axis, we count it as a pass clockwise. On the other hand, if we are processing an edge that starts in the lower halfplane, intersects x^+, and ends in the upper halfplane, we have a counter-clockwise pass.

Note that the same asymmetric reasoning can be used for the original ray casting algorithm. There, it can be phrased as the following rule: every vertex encountered by the ray is considered to be infinitesimally above the ray.

Also note that in the final implementation we can merge the preprocessing step (checking whether the point lies on the polyline) into the actual algorithm. When processing a segment of the polyline, we need to check whether it hits the boy's line of sight. If it hits the boy directly, the answer is true, and if it hits the boy's line of sight we verify whether it counts as a pass in either direction.

In the implementation, it is sufficient to check the signs of the girl's y coordinates to see whether she crossed between halfplanes, and if she did, a single cross product can tell us whether she did so behind the boy, in front of him, or directly through his location. Note that all these checks can be implemented using integers only if the coordinates in the input are all integers. This makes the implementation have an especially low constant factor in its time complexity.

The winding number is just a consequence of the more general Jordan Curve Theorem. It has long been believed [13] that tests based on the winding number are significantly (up to 20 times) slower than the ray casting test. Only recently have there been implementations such as [15] that have shown that the winding number can be computed with the same efficiency. For more background on the winding number and related topics we recommend [9].

3.3.4 Experience

This is another of our metaphors that lends itself nicely to a kinesthetic activity in the classroom. We let the students take a chalk and draw any polygon onto the floor. Then we play out the whole routine described in the Metaphor section. In later rounds, the polygon may be replaced by any closed curve, possibly self-intersecting one. At least once the routine shall be done twice with the same starting conditions, but once with the person playing the boy standing still, the other time with rotating and following the one that walks around the curve.

According to our research, tests based on the winding number are under-represented in textbooks and education—which is something we would like to see change in the future. Our presentation of the metaphor aims to show that they can be as intuitive as the ray casting algorithm, and at the same time easier to implement.

3.3.5 Exercises

3.10. Students sometimes propose the following algorithm to test whether a point lies inside a simple polygon: The point lies inside if and only if we hit the polygon in each direction. (In other words, any half-line that starts at our point has a non-empty intersection with the polygon.) Does this algorithm work? If yes, can it be implemented efficiently?

3.3 Winding Number

3.11. Another way to define the inside of a self-intersecting polyline: A point that does not lie on the polyline is said to be *enclosed* by the polyline if you cannot start at the point and "get to infinity" without crossing the polyline. (More formally: point P is enclosed iff there is no semi-infinite curve disjoint from the polyline that starts at P and contains points that are arbitrarily far from P.)

In Fig. 3.18 the set of enclosed points corresponds exactly to the set of points with a non-zero winding number. Find a polyline for which these two sets differ. Is one of them guaranteed to be a subset of the other?

3.12. Design an algorithm to check whether a point is enclosed by a self-intersecting polyline.

3.13. Assume that the boy stands at $(0, 0)$ and looks in the direction of x^+. Also assume that the girl just walked along a straight line from (x_1, y_1) to (x_2, y_2). Write a short piece of code that will decide whether her walk qualifies as either a clockwise or a counter-clockwise pass in front of the boy.

3.14. Assume that we have a closed polyline with n vertices. What is the largest winding number a point can have with respect to this polyline?

3.4 Polygon Triangulation

3.4.1 Overview

Polygons are a useful geometric structure with many applications in practice. Still, there are situations when a general polygon is too complicated to handle, and we would prefer dealing with simpler geometric primitives only. Luckily, each polygon can be *triangulated*—i.e., divided into a collection of disjoint triangles. A more precise definition follows.

> **Polygon triangulation**
> **Instance:** A polygon with n vertices.
> **Problem:** Find any set of triangles with the following properties:
> – Each vertex of each triangle must be one of the vertices of the polygon.
> – No two triangle interiors share a common point.
> – The union of the triangles is exactly equal to the entire polygon.

By induction, we can easily prove that any triangulation of any polygon with n vertices has exactly $n - 2$ triangles—provided that it exists. In the general case, the existence of a triangulation is far from being obvious.

One thing that is obvious is that the sides of all the triangles are sides and some of the diagonals of the original polygon. Hence, triangulating a polygon can be seen as

Fig. 3.20 A *triangulation* of a *convex polygon*: pick an *arbitrary vertex* and connect it to all *non-adjacent vertices*

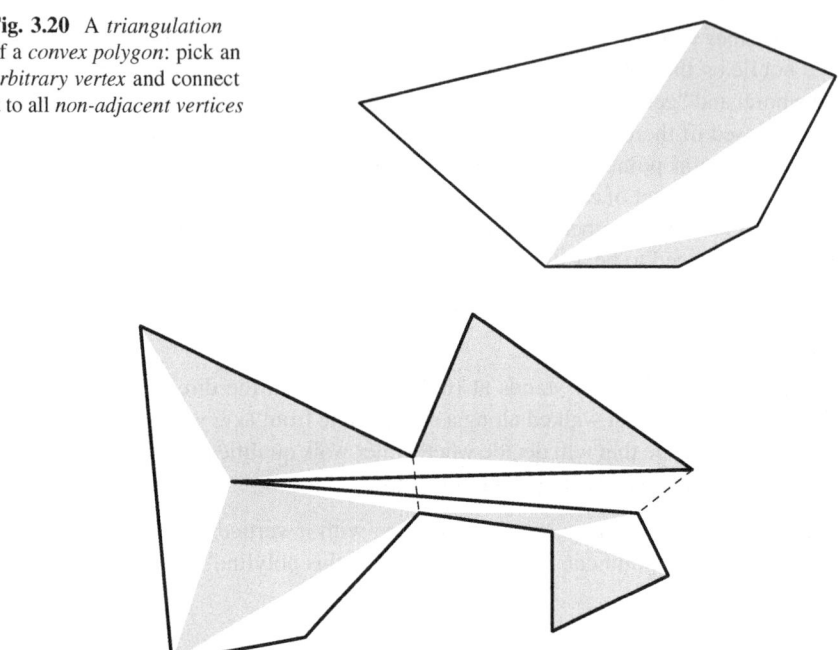

Fig. 3.21 A *triangulation* of a *non-convex polygon*. *Dashed lines* show the *two shortest diagonals*

cutting it into triangles along some of its inner diagonals. (Note that we use the term *diagonal* to denote the line segment connecting any two non-consecutive vertices of a polygon. *Inner diagonals* are diagonals that lie completely inside the polygon.)

Convex polygons are easy to triangulate: Just pick an arbitrary vertex and connect it to all non-adjacent vertices. This set of inner diagonals divides the polygon into the required collection of triangles, as shown in Fig. 3.20.

However, once the polygon is non-convex, the problem becomes more difficult. Obviously, the above algorithm will usually fail, because some of the diagonals will lie partially or completely outside the polygon. Other simple heuristics also fail to construct the triangulation.

For instance, one such heuristic approach would be cutting along the shortest diagonal. (In other words, we want to locate two non-adjacent vertices of the polygon that are closest to each other, and to cut the polygon along the diagonal determined by those two points.) It turns out that in some cases this does not work.

Consider Fig. 3.21. The figure shows a non-convex polygon along with its triangulation. The dashed lines show the two shortest diagonals of the polygon. Note that neither of them is an inner diagonal—one lies completely outside, the other one does even intersect some of the sides of the polygon.

Simple polygons are not the only geometric objects that admit a triangulation. It is also possible (although provably harder) to triangulate any polygon with holes, such as the one in Fig. 3.22. Also, it is possible to triangulate a set of points—find a set of

3.4 Polygon Triangulation

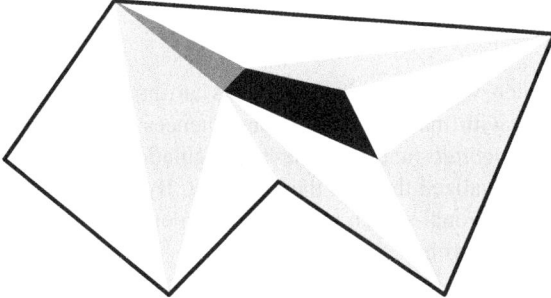

Fig. 3.22 A *triangulation* of a *polygon* with a *black quadrilateral hole*. An interesting thing to note is that we had to use *three shades* to color the *triangles*

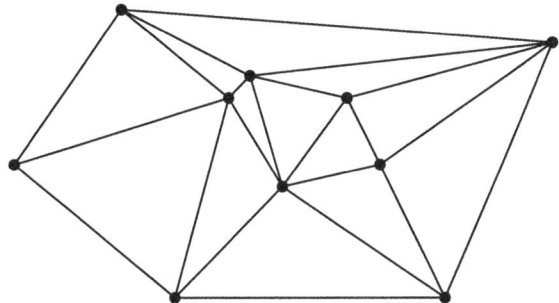

Fig. 3.23 The *Delaunay triangulation* of a *set of points*

non-overlapping triangles with vertices in those points, such that their union is the convex hull of the given points. Among all possible triangulations of a set of points, the Delaunay triangulation [5] has a special place, as it possesses many properties desirable, for instance, in computer-aided modeling (Fig. 3.23). The duality between the Delaunay triangulation and the Voronoi diagram is also often exploited in efficient algorithm design.

Triangulations of geometric objects encapsulate an entire class of important problems of computational geometry with width range of applications in various fields such as computer vision, CAD, motion planning, and many others.

A canonical solution for (possibly non-convex) polygons without holes uses a simple divide-and-conquer technique. As we already saw in Fig. 3.21, the essential part of the algorithm is to find an inner diagonal of the polygon. Such a diagonal is then used to divide the polygon into two independent parts and the algorithm is recursively applied to these two smaller polygons. If we spend $O(n)$ time to find the inner diagonal in a n-vertex polygon, we obtain a solution that runs in time $O(n^2)$.

The metaphor presented below is a constructive proof that every polygon has a triangulation, in that it shows that for $n > 3$ an inner diagonal always exists and that we can find one in $O(n)$ time by following the outlined algorithm.

3.4.2 Metaphor

The metaphor which we explain in here offers an intuitive proof of an involved geometric theorem with many practical consequences: *Every polygon completely contains one of its diagonals* (and hence has a triangulation). The proof is constructive and can be easily visualized through the metaphor. By the metaphor one can get a clear picture of the diagonal-search implementation of a subroutine in the quadratic divide-and-conquer algorithm

As in all the metaphors in this section, we involve a rubber band and another often used concept—the gravity. We describe the metaphor constructively step by step.

First, assemble wooden board to get the boundary of the polygon. The polygon has at least one concave angle. Pick one and fix the board on a vertical wall by a nail through a vertex next to the concave angle. (See Fig. 3.24 for two example polygons.)

Rotate polygon so that both sides of this angle point upward and no edge is horizontal (Fig. 3.25).

Take a lead ball on the end of a rubber band, fix the other end of the band to our vertex with the nail and drop the ball. Clearly, it will fall straight down until it hits a side (Fig. 3.26).

Let the ball slide along the side of the polygon until it reaches a vertex (Fig. 3.27).

At this moment, look at the rubber band. If it is straight, it is the diagonal we are looking for. Otherwise, start at the initial vertex and look for the first obstacle it hits. This has to be some other vertex, and the rubber band piece between them is the diagonal.

3.4.3 Analysis

As we already stated in the Overview section, our metaphor leads to an $O(n^2)$ algorithm to find the triangulation. This algorithm is not optimal. Many textbooks such as [2] also offer a faster (but more involved) $O(n \log n)$ time algorithm.

For a long time, it was an open problem whether polygons without holes can be triangulated faster than in time $\Theta(n \log n)$. The answer was given by Tarjan and van Wyk that discovered an $O(n \log \log n)$ algorithm [16] which was later improved into an $O(n \log^* n)$ algorithm. Later, a linear-time algorithm was discovered by Chazelle [4], but the algorithm is very complex and there have been some concerns about its complete correctness.

Once the polygon may contain holes, the problem becomes more difficult from an algorithmic point of view. Due to a fairly straightforward reduction from sorting, it can be shown that in many computational models the time complexity of triangulating a polygon with polygonal holes has to be $\Omega(n \log n)$.

Finally, we note that there are other known proofs and constructions that lead to an $O(n)$ algorithm to find an inner diagonal. We are not aware of one that would be more intuitive than our original construction that we presented above. Still, there is one

3.4 Polygon Triangulation

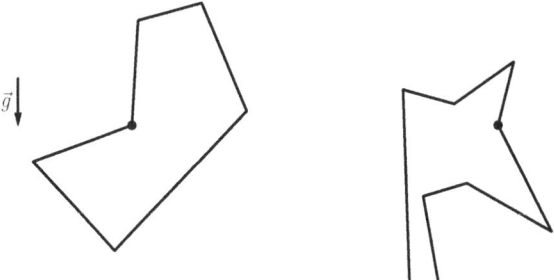

Fig. 3.24 Assemble *wooden board* to get the *boundary* of the *polygon*. Fix it on a *vertical wall* by a *nail* through a *concave vertex*

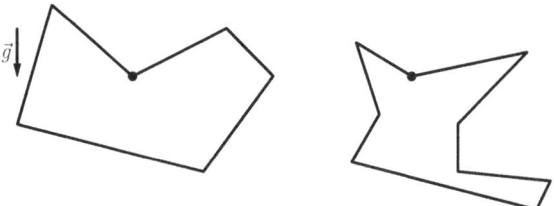

Fig. 3.25 Rotate the *polygon* so that the *edges* from the *nail point upwards* and *no edge* is *horizontal*

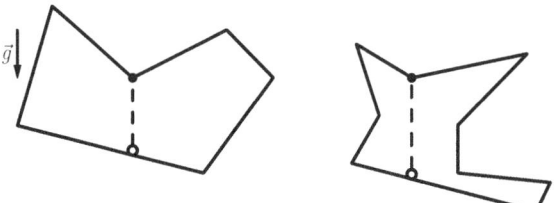

Fig. 3.26 Take a *lead ball* in the *end* of a *rubber band* and fix it in the *other end* of the *band in the nail*. Drop the *ball*. The *ball* falls *straight down* until it hits a *side*

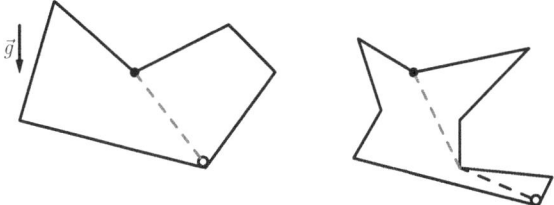

Fig. 3.27 Let the *ball* slide along the *side* of the *polygon* until it reaches a *vertex*

other construction we want to mention due to its simplicity and ease of implementation: Let B be any *convex* vertex of the polygon, and let A and C be its neighbors. If AC is an inner diagonal, we are done. Otherwise, consider all other vertices that lie in the triangle ABC or on its boundary (there have to be some). Let D be the one that is the farthest from the line AC. Then BD has to be an inner diagonal.

One of the reasons why we do not consider the above algorithm intuitive is the fact that the last step *cannot* be replaced by finding the point D' that is the closest to B. (Can you find a counterexample?)

3.4.4 Experience

Polygon triangulation should be considered an intermediate topic, as a good knowledge of the basics of computational geometry is required for any of the algorithms. In particular, dot products and cross products are among the tools necessary to implement any of the algorithms mentioned above.

When presenting our algorithm for triangulation, we make a clear distinction between the algorithm idea (explained via the metaphor) and its implementation. Only once the algorithm is clear do we start discussing its proper implementation: using cross products to find the concave angle, taking the axis of that angle as the direction downwards (instead of actually rotating the polygon), finding the first intersection of that half-line with the polygon boundary, and finally using cross products again to locate the point where the rubber band first reaches the polygon boundary at the end of the process described in the metaphor.

3.4.5 Exercises

3.15. Find some polygons that have a unique triangulation. Is there a polygon with 47 vertices that has a unique triangulation?

3.16. Design a polynomial-time algorithm to recognize whether a polygon has a unique triangulation.

3.17. Show that for a polygon without holes every triangulation is 2-colorable. That is, you can color each of its triangles red or blue in such a way that no two triangles of the same color share a side.

3.18. In Fig. 3.22 we have shown a triangulation of a polygon with holes that is not 2-colorable—or, at least we claim so. Can you prove this?

Are three colors enough for any triangulation of any polygon with holes, or do we sometimes need even more colors?

3.19. Can our metaphor also be used to triangulate a polygon with polygonal holes?

References

1. Asano, T., Asano, T., Guibas, L.J., Hershberger, J., Imai, H.: Visibility of disjoint polygons. Algorithmica **1**(1), 49–63 (1986)
2. de Berg, M., Cheong, O., van Kreveld, M., Overmars, M.: Computational Geometry: Algorithms and Applications, 3rd edn. Springer, Heidelberg (2008)
3. Canny, J.F.: The Complexity of Robot Motion Planning. MIT Press, Cambridge (1988)
4. Chazelle, B.: Triangulating a simple polygon in linear time. Discrete Comput. Geom. **6**(5), 485–524 (1991)
5. Delaunay, B.N.: Sur la sphère vide. Izvestia Akademii Nauk SSSR, Otdelenie Matematicheskikh i Estestvennykh Nauk **7**, 793–800 (1934)
6. Eberly, D.H.: 3D Game Engine Design: A Practical Approach to Real-Time Computer Graphics. CRC Press, Boca Raton (2000)
7. Edelsbrunner, H.: Algorithms in Combinatorial Geometry. Springer, Berlin (1987)
8. Guibas, L.J., Hershberger, J., Leven, D., Sharir, M., Tarjan, R.E.: Linear-time algorithms for visibility and shortest path problems inside triangulated simple polygons. Algorithmica **2**, 209–233 (1987)
9. Haines, E.: Point in polygon strategies. In: Heckbert, P. (ed.) Graphics Gems IV, pp. 24–46. Academic Press, San Diego (1994)
10. Hershberger, J., Suri, S.: Efficient computation of Euclidean shortest paths in the plane. In: Proceedings of the 34th Annual Symposium on Foundations of Computer Science (FOCS 1993), pp. 508–517. IEEE Computer Society (1993)
11. Hershberger, J., Suri, S.: An optimal algorithm for Euclidean shortest paths in the plane. SIAM J. Comput. **28**(6), 2215–2256 (1999)
12. Li, F., Klette, R.: Rubberband algorithms for solving various 2D or 3D shortest path problems. In: Computing: Theory and Applications, 2007. ICCTA '07, pp. 9–19. doi:10.1109/ICCTA.2007.113 (2007)
13. O'Rourke, J.: Computational Geometry in C. Cambridge University Press, Cambridge (1998)
14. Shimrat, M.: Algorithm 112: position of point relative to polygon. Commun. ACM **5**(8), 434 (1962)
15. Sunday, D.: Inclusion of a Point in a Polygon. http://geomalgorithms.com/a03_inclusion.html (2012). Accessed 8 Dec 2012
16. Tarjan, R.E., Wyk, C.J.V.: An $O(n \log \log n)$-time algorithm for triangulating a simple polygon. SIAM J. Comput. **17**(1), 143–178 (1988)
17. Welzl, E.: Constructing the visibility graph for n-line segments in $O(n^2)$ time. Inf. Process. Lett. **20**(4), 167–171 (1985)

Chapter 4
Strings and Sequences

4.1 Stacks and Queues

4.1.1 Overview

The stack and the queue belong among the most basic data structures. They are very closely related to each other. In some sense, these two data structures can be seen as complementary: while the stack represents a LIFO (last in, first out) memory, the queue is a FIFO (first in, first out) type of memory. The main principles of these two data structures are often presented together.

However, there is a significant difference: in the ease of their implementation. While almost any implementation of a stack will have an optimal time complexity (constant time per operation), when implementing a queue for their first time, many beginners will produce an implementation for which the time complexity of some operations turns out to be linear in the number of stored elements.

The difficulty of implementing a queue comes, in part, from the fact that the students who are learning about queues do not possess a firm grasp on complexity theory yet—and therefore they do not realize that the solution they intuitively drafted is in fact extremely slow.

However, the other thing to blame for this problem is inappropriate metaphors used when explaining a queue. These metaphors are what then leads the students to design a bad queue data structure.

In particular, the most popular metaphors used by many textbooks (for instance [11, 15] and many others) are metaphors like:

- The checkout line at a store.
- People waiting on an escalator.
- Cars waiting at a car wash/petrol station.
- Machine parts on an assembly line.

While natural, these are the *worst* possible examples of queuing in terms of operations with the data structure: each time an element is processed, all the others have to move. Some of them are even inefficient in real life. For instance, when a car leaves a car wash, the others all have to start their engines and, one after another, move one spot ahead in the queue. The same goes for the people standing in a checkout line who have to shuffle a few steps ahead each time somebody at the cash register pays and leaves. In the other cases (escalators and assembly lines), the efficiency in real life comes only from the fact that we have a mechanical way of moving all elements ahead at the same time.

4.1.2 Metaphor

We can easily identify many real-life examples of smart queuing, once we start looking for them.

One particular example of smart queues is offices that have a system of numbered tickets. Whenever a client arrives, she has to push a button on a machine. The machine dispenses a ticket with a number. The client then takes a seat and waits until her number is shown on a display board.

Most students are probably familiar with concrete locations (post offices, banks, etc.) that use such systems. Commonly, the tickets use a fixed number of digits, usually three. What happens when the system runs out of numbers? Obviously, it starts again from the beginning: e.g., the next ticket dispensed after the ticket 999 has the number 000 again.

This metaphor directly corresponds to a queue implementation that uses a cyclic array with two pointers: the head of the queue (the ticket number being processed) and the tail of the queue (the next ticket number to print). Clearly, both enqueuing and dequeuing are done in constant time. The metaphor also shares other features with the implementation. For instance, in both cases we have a clear limit on the number of items that can be stored in the queue at the same time.

In many countries, another example of smart queuing can be found in doctors' waiting rooms and other similar places. In such places, people do not stand in line. Instead, they sit wherever they like. And whenever a new patient arrives, she asks a question along the lines of "Who was the last one here so far?" and remembers the particular patient. Now, as soon as that patient is taken into the doctor's office, she knows that she is the next one to go in.

This metaphor loosely (see Exercise 4.1) corresponds to a queue implemented as a linked list. The shared features include operations in constant time, and the potentially unlimited number of items in the queue.

Stacks can also be introduced via a suitable metaphor. Even the simplest metaphors, such as a stack of books or the disks in Towers of Hanoi, work well.

One of the authors of this book has first encountered a stack at the age of 7, when buying ice cream from a street vendor. In Slovakia, it is customary to serve ice cream in scoops placed one atop another, as shown in Fig. 4.1. Of course, the ice cream has to be eaten from top to bottom. On the other hand, the ice cream vendor places the scoops on the cone in the order in which the child asked. The author was then very

Fig. 4.1 An ice cream cone with three scoops of ice cream

pleased with himself after a "discovery" that he can reverse his order before placing it—and then he will get to eat the flavors in the intended order.

Note that this metaphor is strongly culturally biased. It does not work in countries where ice cream is served in a different way. Just explaining it with pictures tends to be useless if students lack the experience. In such cases, the teacher should select a more culturally appropriate metaphor. For instance, we also had a report of a similar experience with a stack of pizza slices bought from a street vendor.

Also note that the ice cream metaphor can be extended to less trivial stack operations, for instance the cone offers "read access" to all its elements, just like a stack stored in an array.

4.1.3 Analysis

We expect that our reader has a good understanding of the scientific background behind stacks and queues, therefore in this case we omit the Analysis section.

4.1.4 Experience

When teaching queues, we like to introduce not only our "smart" queuing examples, but also the traditional ones—but in our case, this is done with the clear intent to show (or to have the students discover) their inefficiency. Kinesthetic activities with smart and naive versions of queuing can also be used to help the students discover the importance of doing all operations in constant time.

For the ticket system metaphor, after introducing it, we recommend asking the students to come up with some answers to the question "What can go wrong with this system?" They should be able to come up with both major problems: having too many people arrive/wait at once, and running out of ticket numbers simply by processing too many people. Usually, the students could also come up with the suggestion how to solve the latter problem. Once they come up with the idea on their own, the queue implementation in a cyclic array will seem natural to them.

Once the students have a clear understanding of efficient and inefficient ways of queuing, we like to give them a "think out of the box" puzzle about improving the efficiency of a checkout line. The solution to that puzzle: instead of moving everyone in the checkout line, just move the cash register to the next person in line.

4.1.5 Exercises

4.1 Can you find one significant difference between a queue implemented as a linked list and the "doctor's waiting room" metaphor mentioned above?

4.2 What other examples of queuing can you identify in real life? Are they efficient or not? Is any of them a good metaphor for the "linked list" implementation of a queue?

4.2 Median as the Optimal Meeting Spot

4.2.1 Overview

In many areas of life, we encounter problems that require us to take a given collection of objects and find a new object that is in some sense at the center of the collection. In statistics, one of the most basic (and most common) operations is estimating the mean value of a set of observations. Another example from this category is the least squares method in regression analysis: for example, we may want to find a line that is the best approximation for a given set of points, given that the quality of the approximation is measured as the sum of square errors.

In machine learning, various approaches to data clustering and classification can be considered to belong to this category.

In applied computational geometry, we may encounter this type of problems as various versions of the facility location problem. In these problems, we are looking for the optimal place for a new facility with respect to a set of given constraints. The quality of a particular location for the new facility is usually measured as a function of distances between the location and existing facilities. For instance, the optimal place for a warehouse that services multiple factories will be the place that minimizes the (appropriately weighted) average distance between the factories and the warehouse.

In this section, we will consider one of the simplest facility location problems: the one-dimensional version defined below.

One-dimensional facility location
Instance: An ordered sequence of points $(x_1, \ldots, x_n)$ on the real axis.
Problem: Find the x that minimizes the sum of distances to all x_i.
Formally, minimize $\sum_i |x - x_i|$.

4.2 Median as the Optimal Meeting Spot

Fig. 4.2 A village with an arbitrarily placed bus stop. The owners of the three houses to the *left* of the bus stop want to pull it farther to the *left*. The other five people pull it to the *right*

4.2.2 Metaphor

We will introduce this facility location problem using a simple story: The real axis is the main road leading through a village. Each of the houses in the village stands next to this road. There is a bus service that would like to service the village. The villagers now have to pick a single location for the bus stop.

Of course, each villager would prefer to have the bus stop next to their house. The farther they have to walk, the unhappier they are about the location. Taking the good of the entire village into consideration, the mayor decided that the location of the bus stop should minimize the *average* distance a person would have to walk from their home to the bus stop.

Before we proceed to the solution, note that minimizing the average distance is exactly the same as minimizing the sum of all distances. This is because the sum can be computed as n times the average, where n (the number of villagers) is a constant. Below, we will talk about minimizing the sum of distances, as the formulas are simpler without the division. (Also, if the inputs are integers, all calculations can easily be done in integers only.)

Now for the solution. What actually happened in the village from our story? On one sunny day, a truck delivered the bus stop sign and dropped it off somewhere close to the beginning of the village. Of course, it did not take long for the entire village to gather at the sign. One word led to another, and soon everyone was arguing. And then one of the villagers grabbed the bus stop sign and started to pull it toward the center of the village. He was quickly joined by everyone else. Soon, the entire village was trying to pull the sign—but not in the same direction. Obviously, everyone was pulling it toward their homes.

An example situation is depicted in Fig. 4.2. For simplicity, from now on we shall assume that each house is the home for exactly one villager.

Of course, as most of the villagers were pulling in the same direction, they overpowered the opposition and started to move the bus stop closer to their homes.

But as they started to pass some houses, more and more of the villagers who were initially pulling as a part of the majority started to change their opinion: first to "hey, this is perfect, how about we stop here?" and immediately afterwards to "stop it, come back with the bus stop!".

What happened in the end? Did they reach some compromise? Or are they still pulling the bus stop back and forth? We shall see that in a minute.

Convergence towards optimality. At this moment, it is important to realize that the process described in the story actually decreases the sum of distances between the bus stop and each of the houses—i.e., the quantity we want to minimize.

Consider a situation in which l people want to pull the bus stop to the left, and r (with $r > l$) want to pull it to the right. The bus stop will start moving in the majority direction. Clearly, nothing will change until the bus stop reaches the next house in its direction. Let d be the distance the bus stop traveled during this phase. When compared to its previous location, the bus stop is now d closer to r houses and d farther from l houses, giving us a total change of $d(l - r) < 0$. That is, the sum of all distances has necessarily decreased.

As long as the number of people on both sides of the bus stop is unbalanced, they will keep on pulling it (and thereby improving the sum of all distances). In other words, the optimal place for the bus stop has to be in a location where the number of people pulling it in each direction will be exactly the same.

Where can we find such locations? Suppose that there are n houses in the village, numbered 1 through n in order. If n is odd, there is precisely one such location. The bus stop has to be placed in front of the house number $(n + 1)/2$, i.e., the house with the *median* coordinate. We can easily verify that if we place it anywhere else, the villagers will pull it toward the median.

If the number of houses happens to be even, there will be multiple locations where the bus stop will be in balance: anywhere between the middle two houses, that is, houses number $n/2$ and $(n + 2)/2$. (And also exactly in front of either of the two houses).

How can we tell which of those locations is the best one for the bus stop? Luckily for us, they all are. This is again easily seen using the same argument as above: moving the bus stop within this segment brings it closer to $n/2$ houses and farther from the other $n/2$ by the same amount. Hence, the sum of distances is the same for any bus stop location within the optimal segment.

4.2.3 Analysis

While the facility location problem on the path graph is easy, many of its generalizations are hard, and only a few of them are solvable in polynomial time. In the generalizations, we can consider other graph classes, vertex weights, more than one facility (this is called the k-median problem if the number of "bus stops" is a fixed constant k), and a different distance metric.

Among the solvable versions are the k-median problem on paths [8] and even on trees [16]. Some special cases of these problems are posed below as Exercises 4.3, 4.4, and 4.6. On the other hand, it has been proved [9] that in general graphs the k-median problem is NP-hard.

In addition to discrete versions of the facility location, there are also continuous versions with practical importance. Among the simplest ones is the version where, instead of minimizing the sum of distances, we are looking for a point on the line

4.2 Median as the Optimal Meeting Spot

that minimizes the sum of squared distances from the given n points. It can easily be shown that the optimal solution in this case is the mean—i.e., the coordinate of the optimal solution is the average of the coordinates of the points given in the input.

In the problem where we minimize the sum of squared distances, the solution extends without any change to multiple dimensions. On the other hand, the median problem (in which we minimize the sum of distances) is hard already in the two-dimensional plane: in [2] it is shown that this problem cannot be solved by radicals over the field of rationals. As a consequence, there is no exact algorithm for this problem in a model of computation where the root of an algebraic equation can be obtained by using arithmetic operations and the extraction of k-th roots. Therefore, the median problem can only be solved by an algorithm that numerically (or symbolically) approximates the solution.

4.2.4 Experience

The metaphor we presented above is very clear and we have been using it without observing any unwanted flaws in the students' thought processes. After seeing the metaphor, the students are usually able to solve many of the exercises presented below independently. We especially recommend Exercise 4.8 where the students need to show both an understanding of the general principle and a careful attention to detail.

4.2.5 Exercises

4.3 Consider the weighted case of the bus stop problem: the input is the number n of houses, and for each house we have its coordinate x_i and its number of inhabitants v_i. We want to find the location of the bus stop that minimizes the average distance a villager has to walk from their home to the bus stop. How can this problem be solved?

4.4 Assume that the topology of the village in our bus stop problem is a general tree, not just a line. For each edge of the tree we are given its length. Can the optimal meeting place still be found efficiently? And what about if we consider the weighted version on a tree (i.e., each vertex of the tree may contain multiple people)?

4.5 There are n people in Manhattan who want to meet (all at once). You know the current coordinates of each person. Compute the meeting place that minimizes the average walking distance. (Assume that they only move by walking, and recall that in Manhattan it is only possible to move in the four cardinal directions: north/south and east/west.)

4.6 A more general version of our problem (with applications in statistics and data mining) is known as the "k-median" problem. In this exercise, we consider its simplest version. As in the original problem, we have a village with n houses built along a road. Place *two* bus stops into the village. For each house, we compute the distance to the *closer* bus stop. Minimize the sum of these distances.

Design an algorithm that will compute the optimal locations of the two bus stops. Try to find a solution with an optimal time complexity.

4.7 The general thought pattern "if a solution can be improved, it is not optimal" is useful in many situations, not just for the median. Solve the following slightly related problem using this method:

There are n kids with buckets of different sizes waiting for a single water pump. In which order should they fill their buckets in order to minimize the average waiting time?

4.8 In the original bus stop problem, let us assume that the mayor added a new constraint: No villager should be forced to walk for more than 1 km in order to reach the bus stop.

First, design an algorithm that will check whether it is possible to satisfy this new constraint at all. Next, design an algorithm that will (assuming that it exists) find the location of the bus stop that minimizes the average walking distance, given the new constraint. That is, out of all bus stop locations that satisfy the new constraint we are looking for the location that minimizes the average distance a person has to walk to reach the bus stop.

4.9 If it turns out that the mayor's new rule (introduced in the previous exercise) cannot be satisfied, the village is so long that it probably should have more than one bus stop.

Design an efficient algorithm that will compute the minimal number of bus stops sufficient to satisfy the mayor's new rule. (You do not have to place the bus stops optimally. I.e., we do not care about the sum of distances people will have to walk, we just want to place as few bus stops as we can.)

4.3 Substring Search

4.3.1 Overview

Locating information is one of the essential aspects of modern society. One of the simplest algorithmic problems related to locating information is the problem of looking for an occurrence of a given shorter string (the pattern, usually called the *needle*) as a contiguous substring of a longer string (the text, usually called the *haystack*).

In the text below, we will denote these two strings N and H. We will also use n and h to represent their lengths. (That is, $n = |N|$ and $h = |H|$.)

4.3 Substring Search

> **Substring search**
> **Instance:** Two strings N and H.
> **Problem:** Check whether N occurs within H as a (contiguous) substring.
> OR: Find one occurrence of N within H, if any.
> OR: Find all occurrences of N within H.

Our kind reader will surely notice that in the above definition box we actually defined three related, but still slightly different versions of the problem. These are different problems and their generalizations have solutions of different efficiency. Luckily, in the simplest case presented above, it turns out that all of these problems can be solved optimally—the time complexity of the solution will be linear in the input size. Clearly, the third problem is the most general one, hence we shall focus on this version. Below, this will be the problem denoted "the substring search problem".

Of course, different applications ask for different solutions to this problem. When somebody uses the built-in search function in their browser or text editor, it is most likely that the internal representation is the brute force one: "for each possible offset x, compare N to H, starting at character x of H". This is perfectly OK, because this solution is simple and sufficient for the task at hand.

However, when the same user later sends a query to a web search engine, a completely different set of algorithms has to come into play. Yet, another chapter is a researcher in bioinformatics who tries to locate a piece of DNA in the gigabytes of available data. Different setting for the substring search problem ask for very different algorithms.

We will focus on one of the earliest (but most versatile and still frequently used) asymptotically optimal algorithm: the Knuth-Morris-Pratt substring search (KMP, [12], also see Chap. 32 in [7]).

One particular issue with the KMP algorithm is that its implementation is notorious for being error-prone.[1] It is especially easy to make off-by-one mistakes in the implementation. Our metaphor also aims to get rid of these mistakes by assigning a clear physical quantity to each state of the algorithm.

4.3.2 Metaphor

In principle, the metaphor will be a mechanical gadget. To a reader well versed in formal languages and automata theory the gadget will certainly resemble a finite automaton. However, there will be some subtle differences between the gadget and the traditional automata (both deterministic and non-deterministic ones).

---

[1] Consider, e.g., http://www.dreamincode.net/forums/topic/273377-knuth-morris-pratt-algorithm%3B-reposted/ and http://pixelstech.net/article/1330941936_Overlap_Detection.

Fig. 4.3 The gadget for the string N = COCOA. There are already some lemmings inside the gadget. The letter C has just been chosen and the corresponding bridges have appeared. The trajectory of each lemming during the next second is shown by the corresponding *arrow*

The Gadget

The gadget will consist of multiple almost identical blocks that will be placed next to each other from left to right. Each block will contain a bottomless pit marked with a single letter. There will be exactly n pits: one for each letter of N (the needle). From left to right, the letters marking the pits will form the string N.

Our metaphor will also use lemmings: the small humanoid creatures made famous by the series of computer puzzle games. The lemmings live by very simple deterministic rules. In our metaphor, the only rule we need is that all lemmings always keep on walking toward the right at a constant rate. Of course, whenever a lemming encounters a pit, it falls into the pit and disappears forever.

Once a second, a lemming appears to the left of the row of pits. Of course, if nothing happens, the lemming walks toward the nearest pit, falls into it and disappears.

We have only one way of preventing that. Once a second we have to say a letter. Once we choose a letter, bridges will magically appear over all pits that are marked by the letter we chose. Thanks to those bridges, some lemmings will avoid falling into a pit. Instead, each of the lucky lemmings will use a bridge to get across the next pit.

Choosing the letter takes us no time, and the bridges also appear in an instant. It takes a lemming precisely one second to cross a bridge. After the second elapses, all bridges disappear and we have to say the next letter.

Figure 4.3 shows a sample gadget with some lemmings.

Getting Lemmings Through the Gadget

Below, the left end and the right end of the gadget will be called its beginning and its end, respectively. Suppose that we really like a particular lemming that just appeared at the beginning of the gadget. We would like to get it to safety—to the house at the end of the gadget. How can we do that? Clearly, there is only one way: while the lemming walks, we have to say exactly the correct sequence of letters: the string N. Each of the letters will activate the bridge our lemming needs during the next second (and possibly some other bridges; we do not care about those at the moment).

4.3 Substring Search

It is also worth repeating that each lemming is deterministic. Hence, whenever there is a lemming at the beginning of the gadget, if we say the letters of N, in order, the lemming will certainly reach the end.

Using the Gadget to Search for Substrings

We shall use our gadget to assist us in looking for occurrences of the given needle N (the one used to build the particular gadget) in a given haystack H. For each letter of H, in order, the following will happen.

1. A new lemming appears at the beginning of the gadget.
2. We read the letter aloud.
3. Some bridges appear.
4. We wait for a second. During that second, the lemmings who have a bridge in front of them walk across a pit, and the other lemmings fall into the pits in front of them and disappear.
5. If a lemming just reached the end of the gadget, we just found an occurrence of N in H.

Figure 4.4 shows what happens each second while we use the gadget for $N = $ COCOA (shown in Fig. 4.3) to look for N in the string $H = $ OCOCOC.

It should be obvious that this process does indeed find all occurrences of N in H: we already know that it works for any single lemming, and now we are just doing it with multiple lemmings in parallel.

Sequential Simulation

Obviously, we do not really care whether the movement of lemmings is continuous—we can easily simulate the process in discrete steps, each corresponding to one second of time. Thanks to this observation, we can directly translate the above procedure into a computer program that will read N and H and then simulate the individual steps.

A good implementation of the simulation stores the set of "active" lemmings. For each lemming, we remember *the number of pits it already crossed*. Whenever a new letter of H is read, we first add a new lemming to the set (the one that just appeared at the beginning of the gadget), and then for each lemming, we check whether it advances or falls into the next pit. This check can be done in constant time—we just compare the current letter of H to the letter of the pit, the lemming is trying to cross. The lemmings that fall into pits are removed from the set of active lemmings.

What is the time complexity of this simulation? The answer is easy: we can simulate each step of each lemming in constant time. Hence, the total number of steps of our algorithm is linear in the number of actual steps taken by all the lemmings.

Also, recall that we already know that simulating a single lemming is equivalent to checking whether N occurs at a particular offset in H. Hence, simulating all lemmings is precisely equivalent to the naive algorithm: for each offset, compare

(a) The first second: The current letter is O, the first lemming falls into a pit.

(b) The second second: The current letter is C, the lemming advances.

(c) The third second: The current letter is O, one lemming advances, the other disappears.

(d) The fourth second: The current letter is C, both lemmings advance.

(e) The fifth second: The current letter is O, two of the lemmings advance.

(f) The sixth second: The current letter is C, the lemming closest to the end disappears.

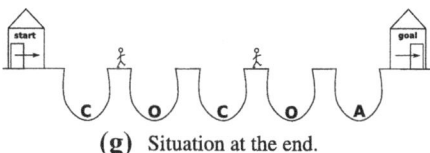
(g) Situation at the end.

Fig. 4.4 Using the gadget for $N = \text{COCOA}$ to process $H = \text{OCOCOC}$

4.3 Substring Search

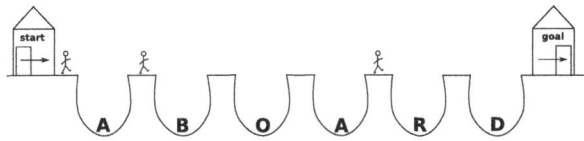

(a) The gadget for the string $N =$ ABOARD. What were the last few letters of H read, given that the lemmings are in their current locations?

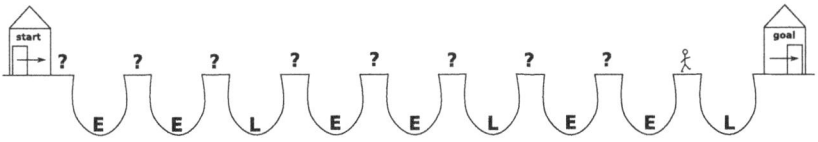

(b) The gadget for the string $N =$ EELEELEEL. Where are the other lemmings?

Fig. 4.5 Sample gadgets that illustrate the information the gadget carries about the past

the letters of N to the letters of the corresponding substring of H, stopping at the first mismatch. (The mismatch corresponds to the pit that claimed that particular lemming.) The only difference between the naive algorithm and the simulation of the gadget is in the order of operations: while simulating the gadget, we are always checking multiple offsets in parallel.

Therefore, we have these two important conclusions:

> We can correctly solve the substring search problem by simulating the gadget with lemmings. A sequential simulation of the entire gadget has the worst-case time complexity of $\Theta(nh)$.

Information About the Recent Past

Before we move on with the exposition, it is helpful to include the following two queries. First, consider the gadget shown in Fig. 4.5a. This figure shows the gadget for $N =$ ABOARD at some point during the simulation. More precisely, this is the moment immediately before the next letter of H is going to be read. The bridges used during the last second have already disappeared. All you can see are the locations of all lemmings.

Can you tell which was the last read letter of H? And can you also tell the one read immediately before the last one? How many most recent letters of H can you reconstruct, based on what you see in the figure?

Second, consider the gadget shown in Fig. 4.5b. As in Fig. 4.5a, this is the state of a gadget after some seconds have elapsed. But this time even the lemmings are hidden from you. All you can see is the string N and the location of the rightmost lemming.

Are there any other lemmings currently inside the gadget? Can you locate them? Are their positions uniquely determined, or are there multiple options? Why is it so?

Here are the answers to the questions in the previous paragraphs. In Fig. 4.5a, we can determine the last read letter of H easily just by looking at the lemmings that are currently inside the gadget. Each of them has just crossed a pit with the letter A, so that had to be the last letter of H read.

Actually, from seeing the rightmost lemming we can deduce even more. We already know that it crossed an A-pit during the last second. But then the second before that the lemming had to cross an O-pit, and so on. In other words: during its lifetime, that lemming has crossed pits that spell ABOA; therefore, those have to be the last four letters of H read before the situation in the figure occurred.

For Fig. 4.5b we can repeat the same deduction to discover that the last eight letters of H had to be EELEELEE.

For each of the positions marked by question marks, there was exactly one lemming that can now be at that position: the one that appeared the right number of seconds ago. Now we need to evaluate whether that lemming actually reached the position (and is still active), or whether it fell into a pit on its way. And for each of those lemmings, the answer can be determined—because we already know the exact letters the lemming heard during its life.

For example, the lemming that might have reached the position immediately to the left of the one we see would hear the sequence ELEELEE. (Those are the last seven letters heard by the rightmost lemming.) On the other hand, the sequence of pits it has to cross is marked by the letters EELEELE. Therefore, for this lemming the second bridge did not appear when it was needed, so this lemming ended its journey in the second pit from the left.

By repeating a similar process for each of the lemmings that started after the one we see, we discover that there are a total of five lemmings in the gadget: the one we see, the one that heard the letters EELEE, and there are lemmings at the three leftmost positions. (These lemmings have heard EE, E, and nothing yet, respectively.)

The entire gadget with everything we deduced is shown in Fig. 4.6. And clearly, the gadget from Fig. 4.5b was not special in any way, we could repeat the same deduction with any other gadget. This brings us to our next important conclusion:

> The location of a lemming can always be used to reconstruct the locations of all other lemmings to its left. In particular, the location of the rightmost lemming can always be used to reconstruct the locations of all other lemmings currently in the gadget. In other words, by forgetting the locations of all lemmings except for the rightmost one, we are not losing any information.

4.3 Substring Search

Fig. 4.6 Reconstruction of the gadget shown in Fig. 4.5b

Speeding up the Simulation

As we already know, at any moment the rightmost lemming carries all the information we have about the current state of the gadget. Also, the rightmost lemming is the closest one to reaching the goal, i.e., completing the traversal of the gadget. If we keep track of it, we never risk missing a match.

In order to avoid off-by-one errors in the following paragraphs, we need to clearly define a way of numbering the positions in the gadget: The *position* of a lemming will be *the number of pits the lemming has already crossed*. (Or, equivalently, the number of seconds the lemming has been alive.) We will use the same number to denote the location where the lemming is standing at that moment. Hence, possible positions of a lemming range from 0 (the beginning) to n (the end of the gadget), inclusive.

In the new, faster simulation we will be keeping track of the position of the currently rightmost lemming in the gadget. Therefore, the *state* of the simulation will be a single integer: *the number of pits the rightmost lemming already crossed*.

Some steps of the new simulation will be faster: if the rightmost lemming gets another bridge, we just move it one step ahead and we are done with processing a letter of H. On the other hand, as soon as the rightmost lemming falls into a pit, we are out of luck. Whenever that happens, we have to check whether and which of the other lemmings in the gadget survive the next step. More precisely, we just need to find the location of the new rightmost lemming.

We already know one way to compute that, but the process we described above is slow. In fact, its worst-case time complexity is $\Theta(n^2)$. This would take the worst-case time complexity of the entire simulation to $\Theta(n^2 h)$—which is worse than the naive algorithm.

Luckily, the information we need does not depend on H, and therefore it can be *precomputed*. More precisely, for each i between 1 and n, inclusive, we shall precompute the answer $S[i]$ to the following question: *If i is the position of the rightmost lemming in the gadget, what is the position $S[i]$ of the second lemming from the right?*[2]

For example, consider the gadget shown in Fig. 4.7. As we can see, the rightmost lemming is at position 8, and that implies that the second lemming from the right is

---

[2] All these values $S[i]$ are well-defined, because there is always a lemming at the beginning. The value $S[0]$ is undefined, as there is no lemming to the left of that one.

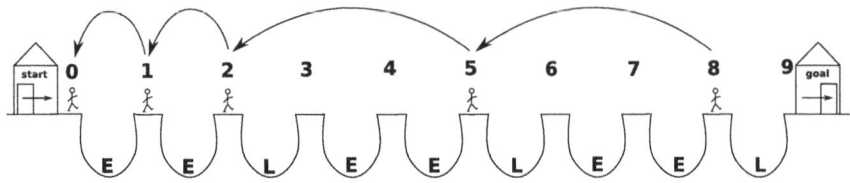

Fig. 4.7 The gadget from Fig. 4.6 with position numbers. The *arrows* represent some of the values stored in S. For example, the *rightmost arrow* illustrates that $S[8] = 5$

at position 5. Hence for this gadget we have $S[8] = 5$. Some of the other values of $S[\cdot]$ for this gadget: $S[7] = 4$, $S[5] = 2$, $S[2] = 1$, and $S[1] = 0$.

In other words, the value $S[i]$ is the index of the next lemming to the left of a lemming who is currently at the position i. And knowing the position of the next lemming to the left actually gives us an efficient way of finding *all* lemmings that are in the gadget at a given time: Let x be the position of the rightmost lemming in the gadget. Then the next one has to be at $S[x]$, the third one from the right (if any) at $S[S[x]]$, and so on, until we reach the lemming at position 0.

For example, in Fig. 4.7 the lemmings are at positions 8, $S[8] = 5$, $S[S[8]] = S[5] = 2$, $S[2] = 1$, and $S[1] = 0$.

Once we have the precomputed array S, we can simulate the entire gadget very efficiently. The new simulation will follow the algorithm given below in pseudocode.

1. Let f be the position of the lemming we currently follow. In the beginning, there is only one active lemming at position 0, thus we initialize f to 0.
2. For each letter x of the string H:

    a. Find the rightmost lemming that survives the step if the next letter is x. This can be implemented as a simple while-cycle:
    While f is defined:
    x. Check whether the lemming at position f survives the next step.
    y. If it does, break the cycle (and then in step 2c increase f by 1).
    z. If it does not, move to the next active lemming by setting f to $S[f]$.
    b. If f is undefined (i.e., no lemming survived), set f to 0: after processing x the only lemming in the gadget will be the one that will appear at the beginning.
    c. Otherwise, increase f by 1: the rightmost lemming who survived moved across the bridge to the next position.

What is the time complexity of this new simulation? Step 2b or 2c is only executed once for each letter of H. But step 2a may sometimes take multiple iterations, sometimes none, so the total time complexity is unclear.

In order to determine the total time complexity, we need to realize that the step 2az is only executed at most once for each lemming—we never go back to a lemming we already stopped following, as that lemming has already disappeared. And the total

4.3 Substring Search

number of lemmings we process is equal to the number of letters in H. Therefore, the total time spent in step 2a is linear in h.

Hence the conclusion for this section:

> If we precompute data that enables us to find the location of the second rightmost lemming in constant time, we can speed up the simulation. Instead of simulating all lemmings, we can just keep track of the currently rightmost one.

Efficient Precomputation

All that remains to conclude our treatment of the metaphor is to explain how the precomputation of the values $S[i]$ is done. The algorithm will look surprisingly familiar: basically, we just saw it in the previous section.

We will compute the values of $S[i]$ sequentially, with increasing i. Suppose that we already know the values $S[1]$ through $S[i]$, and we want to compute $S[i + 1]$. Before we give the general algorithm, consider the example in Fig. 4.8.

In the situation shown in Fig. 4.8, we already know the values $S[1]$ through $S[8]$ (they are shown in italics above the position numbers). With the rightmost lemming at position 8, the second one from the right is at position $S[8] = 5$. The best we can hope for is for that lemming to remain being the second one after the next step. Sadly, we see that it will not happen in our case. The lemming at position 8 will reach position 9 if and only if the next letter is E. And for that letter, the lemming at position 5 will fall into a pit.

As we just lost our most promising candidate, we have to keep on looking. The next active lemming is at position $S[5] = 2$, but if the next letter is E, it will also drop into a pit. The third lemming to consider is at the position $S[2] = 1$, and this lemming will actually survive if the next letter is E. Therefore we have $S[9] = 2$, which is the new position of the lemming from 1 who survived the next step.

Hopefully, it is now clear how this algorithm looks in general. Its implementation is exactly the same as the implementation given in the previous section, with only two differences: Instead of letters of H we will be going through letters of N, and the position f will always point to the second active lemming from the right.

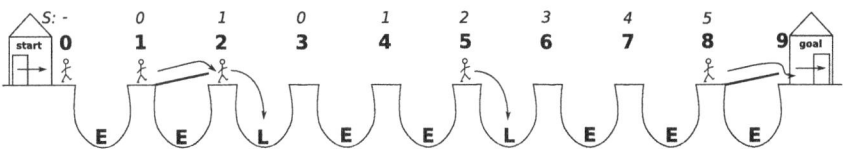

Fig. 4.8 The computation of $S[9]$ for an example gadget. Note that this gadget differs from the one shown in Fig. 4.7: here, the label of the last pit is E

The same argument as before can be used to show that the time complexity of this precomputation is $\Theta(n)$, which makes the total time complexity of the entire algorithm $\Theta(n + h)$.

4.3.3 Analysis

In the recent decades, many versions of the substring search problem were considered and solved. Among the noteworthy discoveries are such versatile data structures as suffix arrays [13] and suffix trees [17, 18]. Both are mostly used in applications where the haystack H is constant (i.e., stays the same during multiple queries).

And even in the simplest setting of a single search there are now many known techniques that all yield efficient algorithms. For instance, the Rabin-Karp algorithm [10] introduces the rolling hash which, since then, has found multiple applications in solving more complex string search problems. The Boyer-Moore algorithm [3] will skip entire portions of the haystack if it is certain that they cannot contain the needle. Still, the Knuth-Morris-Pratt algorithm remains among the "industry standards", in particular due to the amount of information it provides about the string we search.

One particular generalization of the KMP algorithm with many applications (for example in bioinformatics) is its generalization to multiple patterns: the Aho-Corasick algorithm [1].

The time complexity analysis of the KMP algorithm is as involved as the algorithm itself, as it requires amortized analysis. As we have seen above, the metaphor also helps with this aspect—the argument "during the simulation we only start following each lemming at most once" is clear and sufficient to prove the linear time complexity of the whole algorithm.

4.3.4 Experience

During the past years, we have experimented with different variations of our KMP gadget. For instance, we have tried building a vertical gadget consisting of blocks shown in Fig. 4.9. When using the gadget, we are dropping balls into the topmost block at regular intervals, and the gnomes inside the blocks operate the levers to either allow the ball to continue, or to drop it out of the gadget.

The issue with this particular version of the gadget were the switches. In order to simulate the entire gadget, one has to set all the switches correctly after each letter is read. And with this addition, the time complexity of the naive simulation gets worse: it becomes $\Theta(nh)$ for each possible input.

One of the most significant benefits of our metaphor is the huge decrease we observed in the amount of off-by-one errors in the implementation of the KMP algorithm. When we were teaching the textbook version of KMP, our students had made countless mistakes in its implementation. With the clear and tangible definitions

4.3 Substring Search

Fig. 4.9 The building blocks of an alternate KMP gadget

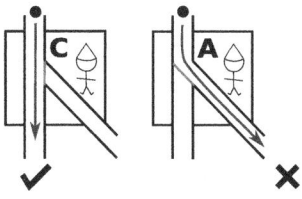

that come from the metaphor (we are following the rightmost lemming, the state is the number of pits it passed, etc.) this is no longer an issue.

Upon understanding the KMP algorithm via the metaphor, our students could easily design various modifications of the algorithm. In particular, some of the students could even independently come up with a major part of the more general Aho-Corasick algorithm.

The way we introduce the problem solved by the Aho-Corasick algorithm is given below as the Exercise 4.13. Note the important feature of the word search problem: no word we search is a substring of another word. A more detailed analysis of the problem, its solution, and our experience with the students' performance is given in the solution to Exercise 4.13.

4.3.5 Exercises

4.10 We have a gadget for some unknown string N. We just finished processing the letter x. There are seven lemmings inside the gadget, counting the one that just appeared at position 0. How many occurrences of the letter x may N contain?

4.11 A string N is periodic with period p if for each valid i we have $N[i] = N[i+p]$, i.e., if the i-th and the $(i+p)$-th letters of N are always equal.

Suppose that you have a gadget for some unknown string N and you already precomputed the values $S[\cdot]$ for this gadget. Can these be used to determine the *smallest* period of N?

4.12 We have a gadget for some unknown string N. The rightmost lemming is currently at position 17 and the second lemming from the right is at position 12. What can we tell about position 7? And what about position 11?

4.13 Given is a word search puzzle: a rectangular grid of letters, and a set of words. The goal of the puzzle is to locate each of the words somewhere in the grid. Each word can be found in one of eight possible locations: horizontally, vertically, or diagonally. It is guaranteed that each word only occurs once in the grid, and that no word is completely overlapped by a different one.

A reasonably efficient solution will use the KMP algorithm to look for each word in each row, each column, and each diagonal of the grid. Try finding a better solution.

Hint: Is it possible to take a row of the grid, read it once and report all the words it contains?

References

1. Aho, A.V., Corasick, M.J.: Efficient string matching: an aid to bibliographic search. Commun. ACM **18**(6), 333–340 (1975)
2. Bajaj, C.: The algebraic degree of geometric optimization problems. Discrete Comput. Geom. **3**, 177–191 (1988). doi:10.1007/BF02187906
3. Boyer, R.S., Moore, J.S.: A fast string searching algorithm. Commun. ACM **20**(10), 762–772 (1977)
4. Canny, J.F.: The Complexity of Robot Motion Planning. MIT Press (1988)
5. Chazelle, B.: Triangulating a simple polygon in linear time. Discrete Comput. Geom. **6**(5), 485–524 (1991)
6. Chen, J., Han, Y.: Shortest paths on a polyhedron, part I: computing shortest paths. Int. J. Comput. Geom. Appl. **6**, 127–144 (1996)
7. Cormen, T.H., Leiserson, C.E., Rivest, R.L., Stein, C.: Introduction to Algorithms, 3rd edn. MIT Press (2009)
8. Hassin, R., Tamir, A.: Improved complexity bounds for location problems on the real line. Oper. Res. Lett. **10**, 395–402 (1991)
9. Kariv, O., Hakimi, S.L.: An algorithmic approach to network location problems, Part II: p-medians. SIAM J. Appl. Math. **37**, 539–560 (1979)
10. Karp, R.M., Rabin, M.O.: Efficient randomized pattern-matching algorithms. IBM J. Res. Dev. **31**(2), 249–260 (1987)
11. Keogh, J.E., Davidson, K.: Data Structures Demystified. McGraw-Hill (2004)
12. Knuth, D.E., James, H., Morris, J., Pratt, V.R.: Fast pattern matching in strings. SIAM J. Comput. **6**(2), 323–350 (1977)
13. Manber, U., Myers, E.: Suffix arrays: a new method for on-line string searches. SIAM J. Comput. **22**(5), 935–948 (1993)
14. Mirzaian, A.: Triangulating Simple Polygons: Pseudo-Triangulations. Tech. rep., York University (1988). Tech. report No. CS-88-12
15. Sedgewick, R., Wayne, K.: Algorithms, 4th edn. Addison-Wesley Professional (2011)
16. Tamir, A.: An $O(pn^2)$ algorithm for the p-median and related problems on tree graphs. Oper. Res. Lett. **19**, 59–64 (1996)
17. Ukkonen, E.: On-line construction of suffix trees. Algorithmica **14**(3), 249–260 (1995)
18. Weiner, P.: Linear pattern matching algorithms. In: Proceedings of the 14th Annual Symposium on Switching and Automata Theory (SWAT 1973), pp. 1–11. IEEE Computer Society (1973)

Solutions to Exercises

Exercises from Sect. 2.1: Paths in Graphs

2.1 In terms of the balls-and-strings model, such a path exists if and only if the string uv has no slack.

In terms of the distances computed by Dijkstra's algorithm, if $\ell(uv)$ is the length of uv, we must have $D[v] = D[u] + \ell(uv)$ or vice versa.

Note that whenever the edge uv satisfies this condition, we can pick t to be either u or v, more precisely the one of those two that is farther from s.

2.2 In terms of the balls-and-strings model, we are asking whether any of the balls drop farther down if we cut the string that represents uv.

If the string uv has some slack, the answer is no. Otherwise, without loss of generality let us assume that v is the deeper of the two balls. Obviously, if there are some balls that would drop after we cut uv, ball v must be among them. Therefore, it is sufficient to check whether this ball drops any farther. And to check that we just need to examine the strings that currently lead from v upwards.

In terms of the distances computed by Dijkstra's algorithm, we need to check whether there is a vertex $x \neq u$ such that $D[x] + \ell(xv) = D[v]$.

2.3 We can easily design graph classes in which the total number of paths between two given vertices grows exponentially in the number of vertices. One such graph class is shown in Fig. 1. Therefore, the best we can hope for in terms of time complexity is a solution that is linear in the output size.

Before actually generating the paths, we will run Dijkstra's algorithm twice: once from s, computing the distances $D_s[\cdot]$, the other time from t, computing the distances $D_t[\cdot]$. Using these two sets of distances, we can easily write a recursive algorithm that uses backtracking to generate all paths from s to t, and nothing else. The main observation that helps us avoid all the other paths: a vertex x lies on some paths from s to t if and only if $D_s[x] + D_t[x]$ equals the shortest distance between s and t (i.e., $D_s[t]$).

2.4 In terms of the balls-and-strings model, note that when the model is hanging by the vertex s, all the edges in all paths from s to t go straight downwards. If we take

Fig. 1 A graph class with an exponential number of paths. (All edges have unit lengths.)

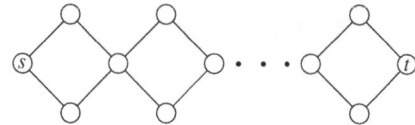

the original graph, leave only those edges, and direct them accordingly, we would get a directed acyclic graph. In Exercise 2.3, we have shown how to generate all paths through such graph using backtracking. Now we need to count those paths.

We can easily add memoization to the algorithm from Exercise 2.3: for each vertex v we will compute and store the number of $s-t$ paths that pass through v— or, equivalently, the number of paths from v to t. When processing a vertex v, we find all its possible successors on the path, recursively find the number of paths for each of them, and sum those counts to get the total number of paths from v to t.

An equivalent solution using dynamic programming starts by computing the distances from s to all other vertices, and then processes the vertices ordered by distance, starting with t and ending with s.

2.5 Let uv be the edge we are going to shorten. In terms of the balls-and-strings model, consider the amount of slack the corresponding string has. If we are going to shorten it by a strictly greater amount, some of the distances will change—the new shorter string will pull the ball on its lower end upwards from its current location. Otherwise, there will clearly be no changes to the graph.

In terms of the distances computed by Dijkstra's algorithm, we need to check whether $|D[u] - D[v]|$ is more than the new length of uv.

2.6 Without loss of generality, let v be the vertex displaced by shortening the edge uv. In the balls-and-strings model, the ball v may pull a set of other balls upwards. We need to discover those balls and recompute the new distances for them. This can be done by marking v as the (currently) only unfinished vertex, resuming Dijkstra's algorithm and marking vertices as unfinished if their distance gets improved.

2.7 Whenever we improve the distance to a vertex, we remember the edge used to do so. When the algorithm terminates, we have a set of $n-1$ remembered edges (one for each vertex except for s). These edges clearly form a tree of paths. The other $m+n-1$ edges can be safely removed. Removing more edges would leave the graph disconnected, thus this number of removed edges has to be optimal.

Fig. 2 A tree with an even diameter, rooted at its center. *Shaded* vertices are the ends of all longest paths. There are exactly 11 distincts longest paths in this tree

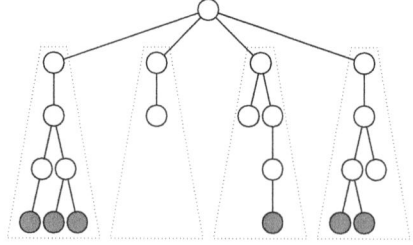

2.8 Fix a particular path. For each string on it: cut it, measure the new depth of t, and then glue the string back. The smallest new depth of t is the length of the second path. This algorithm can be implemented directly, and its running time is clearly polynomial in the number of vertices. (However, note that there are more efficient algorithms for this problem.)

Exercises from Sect. 2.2: Longest Paths in a Tree

2.11 In the described case, it is better to visualize the tree hanging by the center vertex, as shown in Fig. 2. Let d_i be the number of vertices in the i-th subtree that have the globally maximal depth. Each longest path is uniquely determined by picking two of these vertices as its endpoints, with one condition: the chosen two endpoints must not be in the same subtree. The number of longest paths can easily be counted by counting all possible pairs and subtracting the incorrect ones: $\left((\sum d_i)^2 - \sum d_i^2\right)/2$.

2.12 Yes, it does. Exactly the same visualization as in the unweighted version can be used to prove it. (For integer edge lengths you can convince yourself easily by subdividing each edge into edges of length 1 by inserting additional vertices.)

Moreover, it can be implemented in exactly the same way—as the path between any pair of vertices is unique, any tree traversal can be used to compute path lengths and find the farthest vertex.

2.13 No, the algorithm does not work. The improved version given in the next exercise does not work either, we will give a counterexample below.

2.14 On a dense graph, a single breadth-first search takes $\Theta(n^2)$ time. In each iteration, the distance between the current pair of vertices increases. And as the diameter cannot exceed $n-1$, there will be at most $n-1$ executions of breadth-first search, for a total time complexity bound $O(n^3)$.

(At this point, there comes a next exercise: is this upper bound actually tight?)

2.15 A counterexample is shown in Fig. 3. Suppose that the algorithm started in the vertex a. The only vertex at distance 3 is b, all the others are strictly closer. In the next step, we start at b and discover that a is the farthest vertex at distance 3. The algorithm now terminates. However, the conclusion is false—the distance between vertices x and y is obviously 4.

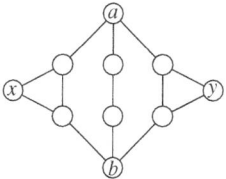

Fig. 3 The tricky longest path algorithm may fail to find the diameter of a general graph

2.16 It is quite common to confuse the diameter and the length of the longest simple path. In trees these two coincide, but in general graphs (even if the edges all have unit length) these are two different concepts. Verifying whether there is a simple path of a given length k is NP-complete—checking for the presence of a Hamiltonian path is a special case of this problem.

On the other hand, the diameter can easily be computed in polynomial time: we can just compute the distance for each pair of vertices, and then output the maximum of these values.

Exercises from Sect. 3.1: 2D Shortest Path

3.1 Yes in both cases. The rubber band metaphor still applies. For a circular obstacle the rubber band will necessarily form a tangent of the circle. Hence, when building the visibility graph, we need to consider tangents from a vertex of a polygon to a given circle, and common tangents of a pair of circles. There is only a constant number of these for any given pair of objects, and they can be found in constant time. See Fig. 4 for an example.

3.2 As suggested by the hint in the statement, there can be situations where the rubber band touches an edge of a polyhedral obstacle. Why is this a problem? Because in such cases we cannot easily tell where exactly the rubber band will touch the edge. In general, there is an uncountable number of possible points where the touch can occur, so the technique of producing a finite visibility graph fails.

However, this should not seen as a shortcoming of our metaphor. Neither should it be seen as just our inability to come up with the right formula to compute the point where the rubber band touches the edge. It has been proved in [15] that in three dimensions the path problem with polyhedral obstacles is NP-hard.

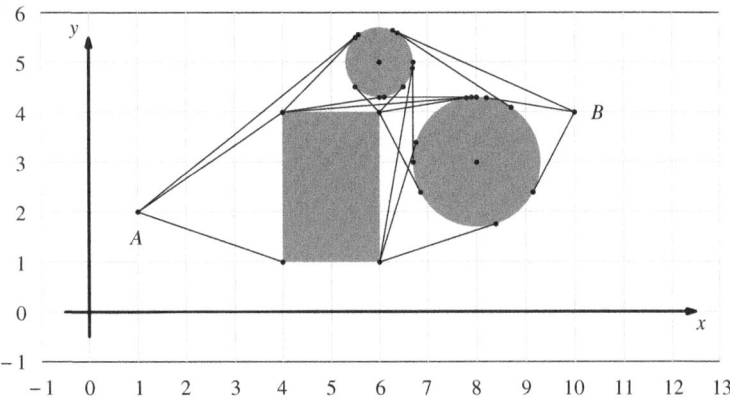

Fig. 4 The visibility graph for polygonal and circular obstacles

Solutions to Exercises

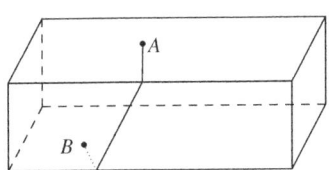

(a) A path from A to B on the surface of a box

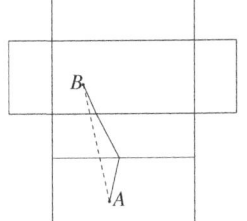

(b) The same box unfolded into a plane, showing the same path (full) and a shorter path (dashed)

Fig. 5 The path problem on two-dimensional surfaces in three dimensions

3.3 The path we seek clearly lies entirely on the surface of the box. Moreover, on each face of the box the path has to be a straight line segment.

Figure 5 shows a sample box and one possible way how it can be unfolded to form a planar surface. Clearly, for the optimal path there must be a way of unfolding the box such that the optimal path turns into a straight line segment that connects the two given points. As for a box there are only a few ways to unfold it, we can try them all and pick the best solution overall.

The unfolding technique was later used in the design of more general algorithms. For example, [16] give a quadratic time algorithm to find the path on the surface of a general (not necessarily convex) polyhedron.

3.4 Basically, the instability occurs around any instance with multiple distinct optimal paths. Consider the instance shown in Fig. 6. In this instance there are two optimal solutions: one goes above the rectangle, the other one below. If we shifted point A slightly upwards or downwards, we would get two very similar instances, each with only one solution.

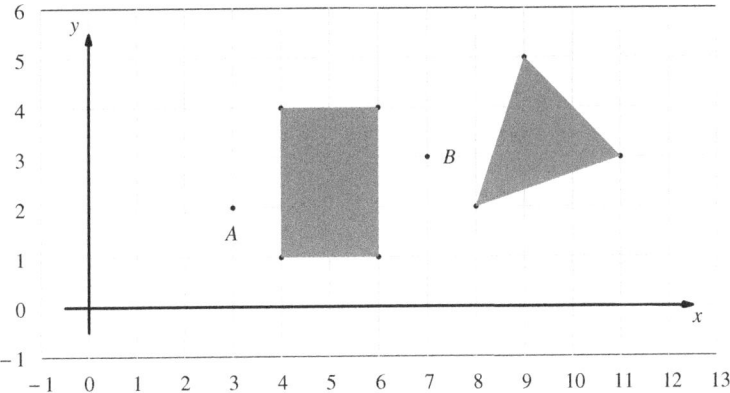

Fig. 6 An instance with two distinct paths

3.5 After we finished the path algorithm on the visibility graph, we can use the computed information to reconstruct all possible paths—or, more precisely for each edge incident to the target vertex we know whether there is a path that uses this edge.

If there is only one such edge (this is usually the case), we have to move B farther away in the direction of the edge. If we leave B in any other direction, the rubber band will be able to contract. Hence it is sufficient to check one position of B. To do this check, we could recompute the path from A to the new B, and compare the distance to the one we expect to find. However, it is sufficient to compute the distances from the new B to the directly visible vertices of the visibility graph. For each of them we already know their distance from A.

If there is more than one optimal direction from which we can reach B in the original graph, the answer is always negative.

Here is an additional exercise. Consider the following simpler algorithm: If all paths reach B using the same edge, move B in the direction of that edge and give a positive answer, otherwise give a negative answer.

Does this algorithm work? Or is the additional check we introduced above really necessary?

Exercises from Sect. 3.2: Distance Between Segments

3.6 Yes, for line segments the two statements are actually equivalent—whenever we have a configuration where neither of the cars moves, nor their distance is not only locally, but also globally minimal. See the answer to the next exercise for more.

3.7 Almost always there is a single configuration where the distance is locally minimal—the globally minimal one. The only situation with more than one local minimum is the one shown in Fig. 3.13 (on p. 41): with two parallel segments there can be multiple optimal configurations.

3.8 For two circles the rubber band metaphor can again be used to easily analyze the possible cases. The railroad car on a circular track will not move if and only if the force pulls it exactly toward the center, or exactly away from it. Hence, if the circles do not intersect, in the optimal configuration the two cars both lie on a line that connects the centers of the two circles. (What happens in the case when the two circles share the same center?)

For the circle-segment case we again get finitely many cases to check: in the optimal configuration, the rubber band has to be orthogonal to the circle, and either orthogonal to the segment, or at its endpoint.

Note that when one of the objects is a circle, in addition to the stable equilibrium that corresponds to the global optimum there is also an unstable equilibrium for each car: the opposite end of the circle.

3.9 The proposed algorithm does not work. The boy in Fig. 3.17 (on p. 47) is outside of the polygon, yet he cannot see the outside in any direction.

Solutions to Exercises

Exercises from Sect. 3.3: Winding Number

3.11 Clearly, all points that have a nonzero winding number are also enclosed by the polyline. However, the opposite inclusion does not have to hold. As shown in Fig. 7, it is possible to have a part of the plane that has a winding number zero but it is still enclosed by the polyline.

3.12 We can easily find the leftmost vertex of the polyline (any of them, if there are multiple). Starting at this vertex, we can walk around the polyline in counterclockwise direction and construct its boundary. (At each self-intersection we change direction to follow the next edge in counterclockwise order.)

The result of this procedure is a new polyline that encloses the same set of points. The new polyline may touch or overlap itself, but it never crosses itself. Hence we can use the winding number or the ray casting algorithm to test whether the given point lies inside this new polyline.

As an example, the boundary of the closed polyline in Fig. 7 is a new closed polyline with 13 vertices.

3.14 For each edge of the polyline, the boy turns by less than $180°$ (half a circle) while the girl traverses that edge. Hence, the number of full circles the girl can make has to be strictly less than $n/2$. For each $n \geq 3$ we can construct a point and a polyline such that the winding number of the point with respect to the polyline will be $\lfloor (n-1)/2 \rfloor$, which is necessarily optimal.

As a simple example, for odd $n \geq 5$, the set of all n longest diagonals of a regular n-gon forms such a polyline.

Exercises from Sect. 3.4: Triangulation

3.15 Among the polygons with the smallest number of vertices, every triangle and every concave polygon has a unique triangulation. For any $n \geq 3$, we can construct an example polygon with a unique triangulation by starting with a triangle and repeatedly adding new triangles such that each new triangle shares and edge with the previously constructed polygon, and the third vertex of the new triangle is placed in such a way that it has no inner diagonal leaving it. See Fig. 8 for a small example.

3.16 Each polygon has a triangulation, therefore each polygon with n vertices has to have at least $n - 3$ inner diagonals. We now claim that the triangulation of a

Fig. 7 Some of the finite regions of the plane may have winding number equal to 0

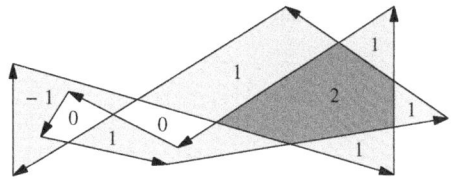

polygon is unique if and only if it has *precisely* $n - 3$ inner diagonals. This is because as soon as a polygon has more than $n - 3$ inner diagonals, it has a triangulation and an inner diagonal d that is unused in that triangulation. If we now use the diagonal d to split the polygon into two smaller ones and triangulate each of them, we will obtain a different triangulation of the same polygon.

The number of inner diagonals can easily be computed in polynomial time. The simplest algorithm runs in $\Theta(n^3)$: for each diagonal and each edge, check whether they intersect.

There are also more efficient algorithms to check whether a polygon has a unique triangulation. Mirzaian [17] shows that it is possible to check whether a given triangulation is unique in $O(n)$ time: it is sufficient to check whether the triangulation contains a diagonal such that the quadrilateral formed by its two incident triangles is convex. Together with Chazelle's $\Theta(n)$ algorithm [18] to find one triangulation this solves our problem in linear time.

3.17 We will construct the coloring recursively. For a triangle the solution is simple, Now suppose we have a polygon with $n > 3$ vertices. As there are $n - 2$ triangles and the polygon has n sides, clearly there has to be a triangle that contains two sides of the polygon. Also, those sides clearly have to be adjacent. If we remove such a triangle, we obtain a polygon with $n - 1$ vertices. We color the smaller polygon recursively. Then we reattach the removed triangle T_1. It shares a diagonal with exactly one other triangle T_2. We look at the color of T_2 and color T_1 using the opposite color.

3.18 Note the triangles labeled 1 through 9 in Fig. 9. Each of these triangles must have a different color from its two neighbors, and that clearly cannot be done with just two colors. In other words, the dual graph is not bipartite, as these nine triangles correspond to an odd-length cycle in the dual graph. Therefore, two colors are not sufficient to color some of the general triangulations.

For an upper bound, we can use the same recursive technique as in Exercise 3.17 to produce a coloring that only uses a few colors. In any triangulation of a polygon with holes there is a triangle that shares at least one side with the boundary of the polygon (including the boundaries of holes, if any). We can look for such a triangle, remove it, and recursively color the rest of the polygon. When we reattach the removed triangle, it becomes adjacent to at most two colored triangles; therefore, we surely have at least one available color for the new triangle.

3.19 Yes, it can. The only exception is the case where the outside polygon is convex, but in that case finding an inner diagonal is easy.

Fig. 8 A 7-vertex polygon with a unique triangulation

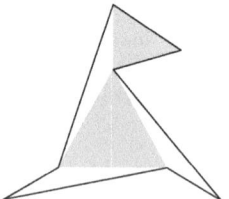

Fig. 9 The proof that *three colors* is required to color this triangulation.

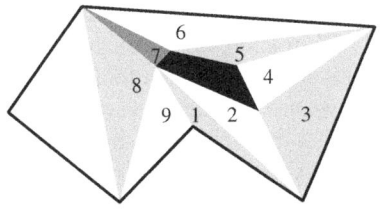

Exercises from Sect. 4.1: Stacks and Queues

4.1 The main difference is in the direction of the "pointers" stored in the data structure. In a linked list implementation of a queue each item is stored along with a pointer to the *next* item, whereas in the waiting room everyone remembers the *previous* patient.

This is a consequence of a more basic difference between the two scenarios: while the items in the linked list are inanimate (and therefore have to be manipulated by an external mechanism), the patients are active agents. From this point of view, the waiting room is a good metaphor for event-driven object oriented programming.

4.2 There are many correct answers to this exercise. One nice example: Each Christmas, I tend to receive some books as presents. And whenever I have a new book, many of my family members and friends are also interested in reading it. This usually evolves into a linked list-like queue: each person remembers an instruction of the type "when you finish reading the book, give it to Deny".

Exercises from Sect. 4.2: Median

4.3 The algorithm works exactly in the same way: we are looking for a location such that there is one half of all people on each side of the location. To prove its correctness, simply assume that each person has their own house. The only difference between this task and the unweighted version is that now there can be multiple houses at the same location.

4.4 Again, we can use the same reasoning. Let v be any vertex of the tree. Assume that we currently have the bus stop in the vertex v. The vertex v has some outgoing edges. There are two possible cases:

(a) The vertex v has a neighbor w such that more than one half of all vertices of the tree lies on the w side of the edge vw. In this case, people living on the w side would manage to pull the bus stop from v to w. In other words, w is a better place for the bus stop, because when moving it from v to w, we are decreasing the walking distance for a majority of people.

(b) The vertex v has no such neighbor. In that case the bus stop will not move from v: regardless of the direction, at least one half of all people will oppose it.Therefore, when looking for the globally optimal location for the bus stop, we are looking for vertices of the second type. The last observation we need is that there can be at most two such vertices in the entire tree, and if there are two, they have to be neighbors. This can easily be shown: if u is a vertex of the second type and v is its neighbor, at most one half of all vertices is on the v side of the edge uv—and therefore at least one half of them is on the u side. Any vertex on the v side other than v will therefore have more than one half of all vertices attached to one of its neighbors.

This gives us a simple greedy algorithm: Place the bus stop into any vertex. While the current vertex is of the first type, move the bus stop to the "large" neighbor. Once this process stops, you have found an optimal location. (The set of all optimal locations either contains this vertex only, or it contains the entire edge leading into one of its neighbors.)

Note that, as in the version on a line, edge lengths do not matter, the optimal location of the bus stop is determined by the shape of the tree only. Also note that the same reasoning can be used to solve the weighted version, we will just use the sum of vertex weights (i.e., the number of people) instead of the count of vertices.

This algorithm can be implemented in $O(n)$ if we first run a depth-first search on the tree to precompute subtree sizes. There is also a different algorithm that solves the problem in $O(n)$: during the first depth-first search we can also compute the sum of walking distances if the bus stop happens to be in the root of the depth-first search. Then, we run a second depth-first search from the same vertex. During this search we take the bus stop with us and update the sum of walking distances as we move along the graph.

4.5 The key to solving this task is to realize that we can break it down into two independent one-dimensional problems. Any solution has to consist of two independent parts: The result of north/south movements of all people has to be that they all end up on the same latitude. At the same time, all their east/west movements have to bring them to the same longitude.

Hence, the optimal meeting place can be determined by computing their median latitude and longitude. (For odd n, there will be a single optimal meeting place. For even n there may be a rectangle of optimal meeting places.)

4.6 Clearly, in the optimal solution there will be some k such that the people from the first k houses in the village use the first bus stop, and people from the other $n-k$ houses use the second bus stop. We can try all possibilities for k and pick the best one. Given k, we can find the optimal placement for each bus stop separately using the original algorithm. This can be done in constant time, but then we have to spend linear time to compute the sum of distances the people who use the bus stop have to walk. (We need this in order to compare the solutions for different k.) The time complexity of this algorithm is $\Theta(n^2)$.

Solutions to Exercises

The above solution can be improved to $\Theta(n)$. For each bus stop we will store the current sum of distances. As we change k, we simulate moving the bus stops and change the sum of distances accordingly.

4.7 Consider any two kids that fill their buckets immediately after one another. If the second kid has a smaller bucket than the first one, we can swap them. (The sum of their waiting times decreases, and none of the other waiting times is influenced by this change.)

Hence, if the sequence in which the kids fill their buckets is *not* sorted according to bucket size, the sum of all waiting times is *not* optimal. This leaves us with just a single candidate for the optimal order.

4.8 Consider the distance between the first and the last house in the village. If this distance exceeds 2 km, it is obvious that the bus stop cannot be within 1 km from each of them at the same time. On the other hand, if the distance between these two houses is 2 km or less, we can find a valid placement for the bus stop by placing it halfway between the first and the last house in the village. (This placement does satisfy the 1 km constraint for each house, but will usually turn out to be worse than optimal in terms of average walking distance.)

Let us take a closer look at how the new constraint limits the possible choices for the bus stop location. For each house, we have a 2 km long interval that has to contain the bus stop. The intersection of all these intervals is the set of all valid locations for the bus stop. Clearly, the intersection of all those intervals is again an interval, and we can compute it as the intersection of just two intervals: the one for the first house and the one for the last house.

If there is an optimal location of the bus stop (computed without the 1 km constraint) that lies within the allowed interval, we are lucky—we can choose that location and be certain that our choice is optimal.

The other possibility is shown in Fig. 10. As the globally optimal solution (i.e., the median coordinate) lies outside of the allowed interval, we can make the following conclusion: If we place the bus stop at any allowed location, there will always be more houses on one side than the other. Hence, the villagers will always want to move the bus stop closer toward the median. Therefore, clearly the best allowed location is at that endpoint of the allowed interval that is the closest to the median.

4.9 There is an efficient greedy solution. Order the houses according to their coordinate. Consider the leftmost house in the village. This house must have a bus

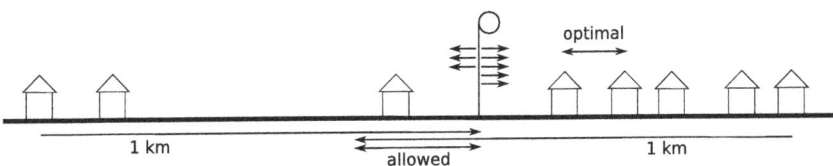

Fig. 10 The additional constraint on 1 km proximity limits the possible locations of the bus stop to an interval. If the globally optimal bus stop location happens to be outside that interval, the optimal allowed solution is at the closer endpoint of the allowed interval

stop within 1 km. Clearly, the optimal place for that bus stop is exactly 1 km in the direction in which the rest of the village lies—no other placement can "cover" more other houses. We repeat the process (pick the first house that is not covered, place a bus stop 1 km behind it) until no more houses remains.

This algorithm can be implemented with time complexity $O(n \log n)$. The slowest part is the sorting; the covering phase can be done in linear time.

Exercises from Sect. 4.3: Substring Search

4.10 Six of the seven lemmings have survived the previous step. Therefore, there must be at least six different pits labeled x in the gadget. The conclusion: N contains at least six occurrences of the letter x. (Note that it may well contain more than 6 xs.)

4.11 The answer is $n - S[n]$. In words: the smallest period is the distance between the two rightmost lemmings in the gadget at the moment when the rightmost lemming reached the end of the gadget.

We are looking for the smallest p such that for each valid i we have $N[i] = N[i+p]$. In other words: Let $k = n - p$. We are looking for the largest k such that the string formed by the *first* k letters of N is the same as the string formed by the *last* k letters of N. (See Fig. 11.)

Suppose that we used the gadget for the string N, and that we read the string N to get one of the lemmings to the end of the gadget. Let us take a look at the position k. Does this position currently contain a lemming? During its lifetime the corresponding lemming had to cross the first k pits in the gadget. On the other hand, during its lifetime the corresponding lemming would hear the *last* k letters of N. Therefore, position k contains a lemming if and only if the first k letters of N are the same as the last k letters.

Hence, the position of each lemming defines one possible period. As we want the shortest period, i.e., the largest k, we are interested in the second lemming from the right. And that lemming is located at the position $k = S[n]$.

4.12 Let $N[x..y]$ denote the substring formed by the letters x through y of N. We know that the rightmost two lemmings are at positions 17 and 12, therefore (by the same argument as in the solution of Exercise 4.11) we have $N[1..12] = N[6..17]$. This means that the shortest period of the string $N[1..17]$ is $p = 5$. But then the string also must have the period $2p = 10$, therefore $N[1..7] = N[11..17]$. And from this equality it follows that there will certainly be a lemming at position 7.

```
abracad abra
 ◄──p──► abra cadabra
         k
```

Fig. 11 The shortest period p is determined by the longest k such that the prefix and suffix of length k match

Fig. 12 A schematic view of the Aho-Corasick gadget for the words *IGUANA*, *LION*, *PEACOCK*, and *PELICAN*. From the position of the lemmings we can infer the last four letters read: *PELI*

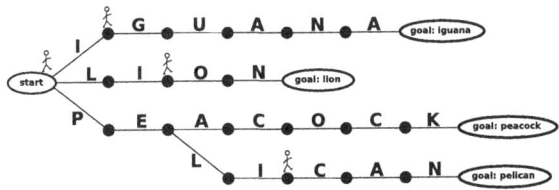

Now suppose there is a lemming at position 11. As there is another one at position 12, it follows that $N[1..12]$ is periodic with period 1. In other words, $N[1..12]$ consists of 12 copies of the same letter. Then, from $N[1..12] = N[6..17]$ we get that $N[1..17]$ has to consist of 17 copies of that letter. But that contradicts the fact that the lemming at position 12 is the second one from the right. Therefore, there cannot be a lemming at position 11.

4.13 We can extend the KMP gadget into one that can look for multiple patterns at the same time. The main trick is to change the single path into a tree of paths. More precisely, the layout of the gadget will directly correspond to a trie that contains all the words we wish to find. The behavior of lemmings will still be deterministic, we just add a single new rule: If you have multiple options how to go on, and one of them currently has a bridge, pick that option.

Figure 12 shows, in a schematic way, a gadget that can be used to look for four words at the same time.

In order to find all occurrences of all words within a given string H, we can use the same process as before: we read the individual letters of H while following the rightmost lemming. (Note that only one lemming is released each second, therefore there cannot be two lemmings tied for being the rightmost one—or for any other position.)

The precomputed information now has to be generalized as well: now we need to answer the question: "If the rightmost lemming is in this note of the trie, in which node is the second lemming from the right?" Note that the answer node may be in a different branch from the question node—as is the case in Fig. 12.

The construction above is the basis of the Aho-Corasick algorithm.

Note that in the general case there is one more complication we did not have to deal with. In general, it is possible that our set of patterns contains two patterns such that one of them is contained in the other. For example, if we are looking for the strings *BAT* and *ALBATROSS* in the string *ALBAT*, we will follow the rightmost lemming on its way through the "albatross branch" of the gadget, completely missing the fact that one of the other lemmings reached the goal in the "bat branch". In the Aho-Corasick algorithm this is handled by adding output links; we omit the details here.

Index

A
Abstraction, 3, 4
Algorithm
 Aho-Corasick, 76, 77, 91
 angle sum, 47, 48
 Bellman-Ford, 18
 Borůvka, 20
 Boyer-Moore, 76
 Dijkstra, 11–20, 34, 78, 79, 80
 divide-and-conquer, 53, 54
 Jarník, 20
 Knuth-Morris-Pratt, 67, 76
 longest path, 24, 27, 28, 81
 Rabin-Karp, 76
 ray casting, 45, 48–50, 85
 winding number, 46
Analogy, 1–6
Anthropomorphic metaphor, 2
Arctangent, 49

B
Backtracking, 79, 80

C
Cayley's formula, 20
Convex hull, 53
Cross product, 50, 56
Curve, 31, 33, 34, 50
 semi-infinite, 51

D
Diagonal, 52–54, 85, 86
 inner, 52–54, 56, 85
Dot product, 43, 56

Dynamic programming, 18, 27, 80

E
Euclidean distance, 35, 42

F
FIFO, 59

G
Geometry
 computational, 31, 37, 38, 46, 53, 62
 multi-dimensional, 43
 three-dimensional, 43
 two-dimensional, 31, 33, 34, 37, 40, 45, 83

H
Halfplane, 49, 50

K
Kinesthetic activity, 36, 50

L
LIFO, 59
Line segment, 32–34, 37–47, 49, 52, 83, 84
Local optimization, 34
Longest path, 20–29, 80, 81

M
Median, 62–66, 87–89
Mental model, 4

Metaphor, 1–8
  Ali Baba's cave, 5
  ball, 13–18, 26, 76, 79
  ball of string, 79, 80
  balls-and-strings, 13–17, 20, 21, 26, 79, 80
  bus stop, 63–66, 87–90
  checkout line, 8
  doctor's waiting room, 62
  gravity, 13–15, 22, 26, 54
  ice cream cone, 61
  inflated balloon, 47
  labeled box, 4
  lemming, 68–78, 89–91
  matryoshkas, 8
  piggy bank, 5
  potential energy, 5
  railroad car, 83
  railroad track, 39–41
  rubber band, 32, 33, 36, 39–41, 44, 54, 55, 81–83
  stack of books, 60
  string, 13–17, 78, 79
  ticket system, 60, 61
  Towers of Hanoi, 60
Motion planning, 31

N
Negative cycle, 12
NP-complete, 28, 82
NP-hard, 36, 82

O
Obstacle, 31–37, 54, 82
  circular, 36, 82, 84
  non-polygonal, 34
  polygonal, 31–34, 36, 37, 82
  polyhedral, 36, 82
Optimization problem, 31, 78
Orthogonal, 40–42, 84
Orthogonal projection, 39, 41, 42

P
Personification, 2
Plane
  two-dimensional, 31, 33, 36, 37, 45, 83
Polygon, 31–34, 45–48, 51–57, 82, 84–86
  convex, 52

  non-convex, 52, 53
  simple, 53
Polyhedron, 37, 83
Polyline, 33–35, 45, 48–51, 85
  closed, 45, 48, 49, 51, 84
  self-intersecting, 48, 51

Q
Queue, 8, 59–62, 87

R
Rubik's cube, 11

S
Search
  breadth-first, 2, 18, 26, 28, 81
  depth-first, 26, 88
Shortest path
  geometry, 31–38, 40, 62
  graph, 11–14, 17–20, 34, 37, 78, 79, 80, 84, 88
  three-dimensional, 37, 83
  two-dimensional, 31, 32, 34, 83
Spiral, 35
Stack, 59–61, 87
Sweep
  plane, 36
  polar, 34

T
Tangent, 35, 82
Topological sort, 2, 18
Tournament graph, 3
Triangulation, 47, 51–54, 56, 85, 86
  Delaunay, 53, 57

V
Visibility graph, 34–37, 82, 84
Voronoi diagram, 53

W
Winding number, 45, 48–51, 85

The manufacturer's authorised representative in the EU is Springer Nature Customer Service Centre GmbH, Europaplatz 3, 69115 Heidelberg, Germany. If you have any concerns regarding our products, please contact ProductSafety@springernature.com

Printed and bound by CPI Group (UK) Ltd, Croydon, CR0 4YY

27/03/2026

02080143-0001